$$\{H\} = [I]\{\omega\}$$

$$T = \tfrac{1}{2}\,\omega \cdot \mathbf{H}$$

APPLIED MECHANICS:
MORE DYNAMICS

BOARD OF ADVISORS, ENGINEERING

APPLIED MECHANICS:
MORE DYNAMICS

CHARLES E. SMITH

OREGON STATE UNIVERSITY

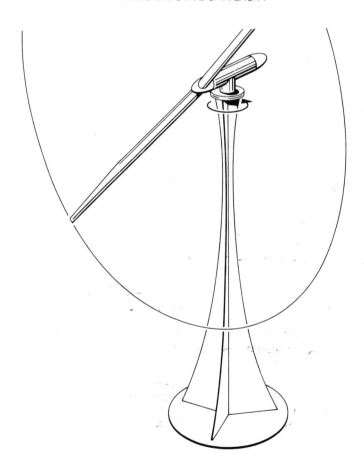

JOHN WILEY & SONS, INC.

New York   London   Sydney   Toronto

Library of Congress Cataloging in Publication Data:

Smith, Charles Edward, 1932—
  Applied mechanics.

  Includes bibliographies and indexes.
  CONTENTS: v. 1. Statics.—v. 2. Dynamics.—
v. 3. More dynamics.
  1. Mechanics, Applied. I. Title.

TA350.S57 620.1 75-44021
ISBN 0-471-79996-3 (v. 3)

Printed in the United States of America

10 9 8 7 6 5 4 3 2 1

PREFACE

Chapters 14 and 16 of this volume present generalizations of kinematics and rigid body dynamics begun in the previous volume, *Dynamics*. Chapter 15 covers matrix algebra and coordinate transformations, for use in curricula that do not include this material. Chapter 17 presents generalized coordinates and Lagrangian dynamics, a generalization of the virtual work introduced in *Statics*. Chapter 18 deals with the dynamic behavior of systems, with properties of linear, constant-coefficient models given in detail.

Chapters 14, 15, and 16 can be used for a course of approximately three or four quarter hours on rigid body dynamics. Chapters 15 and 18 can be used independently of the other three for a course of about three or four quarter hours on system dynamics. Such a course provides a good foundation for further study of vibrations or control systems. This entire book can be covered in an academic year at the rate of about three credits per term.

Chapters 9, 10, and 11 (exclusive of 11-4) and Sections 12-1 of and 12-2 of *Dynamics*, contain the necessary background for a course in rigid body dynamics. In addition, Chapter 13 is strongly recommended. Chapters 17 and 18 can be readily understood with background from any good first-year course in statics and dynamics. A few examples in Chapter 18 require rudimentary knowledge of electrical circuit elements and electromechanical interaction.

In a slight deviation from the usual development of rigid body kinematics, rigid body *motion* kinematics is developed prior to the analysis of rigid body *displacements*. The connection is pointed out near the end of the discussion of displacements. An obvious pedagogical advantage is the treatment of the simpler topic first. Also, this arrangement makes it possible to skip the relatively involved subject of displacements where, for a particular group of students, rotational motion and other subjects that fill the available time are judged to be more important.

The stress in understanding mechanics in geometric terms continues through Chapter 16. As stated in the prefaces to *Statics* and *Dynamics*, I favor concentration on the geometric interpretation of vector relationships over the emphasis so often given to analysis in terms of sets of components. An example is the common development of Equation 14-1:

$$\mathbf{r}\ x\mathbf{i} + y\mathbf{j} + z\mathbf{k}$$
$$\dot{\mathbf{r}} = \dot{x}\mathbf{i} + \dot{y}\mathbf{j} + \dot{z}\mathbf{k} + x\dot{\mathbf{i}} + y\dot{\mathbf{j}} + z\dot{\mathbf{k}}$$
$$= \dot{x}\mathbf{i} + \dot{y}\mathbf{j} + \dot{z}\mathbf{k} + x(\Omega \times \mathbf{i}) + y(\Omega \times \mathbf{j}) + z(\Omega \times \mathbf{k})$$
$$= \dot{x}\mathbf{i} + \dot{y}\mathbf{j} + \dot{z}\mathbf{k} + \Omega \times (x\mathbf{i} + y\mathbf{j} + z\mathbf{k})$$
$$= (\dot{\mathbf{r}})_{\text{rel}} + \Omega \times \mathbf{r}$$

The alternative presented by some authors and in Chapter 14, requires more thought from the student than does the above, but it is not until such thought is given that the relationship is understood well enough for successful application. Another example is the use of Euler's equations (a particular coordinate expansion of the moment-angular momentum relationship) as the basis for working problems of rigid body kinetics. A better understanding results if the basis for analysis is a picture showing the angular momentum resultant and the manner in which it varies with time.

The notations used to distinguish reference frames may seem peculiar to some, but hve been adopted for good reasons. Patterned after the scheme used by Kane*, the subscript preceding the letter indicating a velocity, acceleration, or another vector has proved to be an effective reminder of the reference frame from which the quantity is reckoned. The small Greek letter above a vector symbol denotes the time derivative as observed from the indicated reference frame. Although used by relatively few authors, this notation conveys meaning much more clearly than the more common dot (for time derivative) and subscript (for reference frame). For example, $_\alpha\overset{\beta}{\mathbf{H}}$ says for itself "the β-observed derivative of the α-observed angular momentum," whereas a more traditional $(\dot{\mathbf{H}})_\beta$ requires a lengthy explanation to clarify the roles of the two reference frames.

CHARLES E. SMITH

Corvallis, Oregon
July, 1975

*T. R. Kane, *Dynamics*, Holt, Rinehart, and Winston, 1968.

ACKNOWLEDGMENTS

There were many contributors to this project, and I thank all of them. Valuable suggestions came from many students, especially Brad Whiting and John Gale. Dr. Hans J. Dahlke pointed out a number of errors and shortcomings in *Statics*. Dr. William E. Holley, Dr. Robert W. Thresher, and Dr. Robert E. Wilson contributed significantly to *Dynamics*. I am grateful to the staff of Wiley for the editing and production. The illustrations were prepared by John Balbalis. The most eminent source of inspiration, is Emeritus Professor Kenneth E. Bisshopp of Rensselaer Polytechnic Institute. Finally, for the time made available, I thank Marian, Brian, and Susan.

C. E. S.

CONTENTS

14 Rotational Motion   **1**

   14-1 Derivatives of a Vector   1

   14-2 Particle Kinematics   5
            General Relationships for Velocity and Acceleration   9
                Coriolis Component of Acceleration   12

   14-3 Particle Kinetics in Rotating Reference Frames   23

   14-4 General Motion of a Rigid Object   29
            The Angular Velocity Vector   29
            Composition of Angular Velocities   35
            Screw Motion   40

15 Coordinate and Vector Transformations   **49**

   15-1 Some Matrix Algebra   49
            Definitions   49
            Products of Vectors   54

15-2 Rigid Rotation of Axes 57
The Inverse Rotation 62

15-3 Derivatives of a Vector in Matrix Notation 63

15-4 Linear Vector Transformations 67
Components of a Vector Operator in Different Coordinate
Systems 69
Principal Directions 71

16 Rigid Body Dynamics 77
16-1 Displacement Kinematics 77
Euler's Theorem on Rigid Rotations 79
Chasle's Theorem 80
Screw Displacement 80
Matrix Analysis for Rotational Displacements 81
Direction Cosines and the Single, Fixed-Axis Rotation 83
Infinitesimal Rotations 88
Euler's Angles 91

16-2 Angular Momentum and Kinetic Energy 98
The Inertia Tensor 100
Parallel Shift of Coordinate Axes 102
Principal Axes of Inertia 103
Kinetic Energy of Rotation 106
Cauchy's Inertia Ellipsoid 107

16-3 Application of $\mathbf{M} = \dot{\mathbf{H}}$ 114

16-4 The Work-Kinetic Energy Integral for a Rigid Body 124

16-5 Some Analytically Solved Problems 126
Torque-Free Motion 127
The Spinning Top 130

17 Virtual Work 135
17-1 Degrees of Freedom and Generalized Coordinates 135

17-2 Generalized Force Components 142
Conservative Forces 144
Friction Forces 146

17-3 Lagrange's Equations 149

17-4 Integrals of Lagrange's Equations 152
The Lagrangian Function 152
The Hamiltonian Function 152
An Energy Integral 154

Generalized Momentum   154
Ignorable Coordinates   154
The Spinning Top Again   155

18 Dynamic Behavior of Systems   **159**
18-1 A Sampling of Physical Systems   159
      A Hydraulic Servoactuator   160
      A Single-Degree-of-Freedom Oscillator   162
      An Electromechanical Shaker   163
      A Two-Degree-of-Freedom Structure   165
      An Infinite-Degree-of-Freedom Structure   166
      Recapitulation   167

18-2 Classification of Systems   170
      Lumped and Distributed Parameters; Order of Systems   170
      Linear and Nonlinear Systems   171
      Autonomous and Nonautonomous Systems   172
      Various Equivalent Forms of Equations of Motion   172

18-3 Linearization   175

18-4 Implications of Linearity: Superposition   180

18-5 Response of Linear, Constant Parameter Systems   185
      "Natural" Motions   185
            First-Order Systems   185
            Second-Order Systems   186
            Damping Ratio   187
            A Third-Order System   190
            nth-Order Systems   193
            Complex Roots   194
      Forced Responses   197
            Construction of Some Particular Solutions   197
            Input and Output   200
            Steady-State Frequency Response   201
            Indicial Response   203
            Impulse Response   205
            Superposition Integrals   210

18-6 Some Phenomena Peculiar to Nonlinear and Parametrically
            Excited Systems   215
      Multiple Equilibrium Points   216
      Self-Excited Oscillations   216
      Jump Phenomenon   217
      Systems with Periodic Parametric Excitation   218

APPENDIX A   Some Useful Numerical Values   221

    A-1   Physical Constants   221
    A-2   Prefixes for SI Units   222
    A-3   Units of Measurement   222

APPENDIX B   Some Formulas of Vector Analysis   224

    B-1   Definitions   224
    B-2   Formulas   226

APPENDIX C   Some Formulas of Matrix Algebra   227

    C-1   Definitions   227
    C-2   Formulas   228

APPENDIX D   Properties of Lines, Areas, Volumes, and Solids   229

    D-1   Lines   229
    D-2   Plane Areas   230
    D-3   Volumes   232
    D-4   Second Moments of Mass of some Homogeneous Bodies   233

APPENDIX E   Some Relationships Among Complex Numbers   236

    E-1   Definitions   236
    E-2   Useful Relationships   237

REFERENCES   239

INDEX   241

ROTATIONAL MOTION

There are many problems in dynamics that require an understanding of how the observation of motion from one reference frame is related to the observation of the same motion from another reference frame that is moving with respect to the first. Figure 14-1 illustrates an example: the "distortion" of the plane circular motion of an orbiting satellite caused by the earth's rotation about its polar axis. Many other examples are found in mechanisms where the motion of one part relative to another may be determined relatively easily, but be quite complex as observed from an inertial reference frame. The analytical instruments for handling these kinds of kinematics problems will be developed in this chapter.

14-1

DERIVATIVES OF A VECTOR. We conceive a *reference frame* as a set of points in space, the distance between every pair of points remaining constant. At least four noncoplanar points are required. With the introduction of a suitable coordinate system in the reference frame, the two end points defining a vector may be located and a mathematical description of the vector given in terms of a set of components.

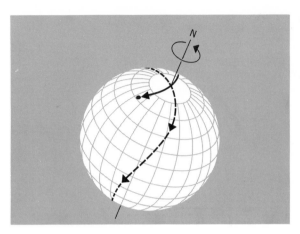

Figure 14-1

In this section we examine the effect that the motion of one reference frame relative to another has on the time derivative of a vector.

Consider the appearance of a vector **A** from two different reference frames, frame β rotating with respect to frame α. Suppose that the vector appears to be constant in frame β, so that the β-observed derivative of **A** is zero. Then, from frame α, **A** will appear to be changing its direction, so that the α-observed derivative of **A** is *not* zero. Thus, starting with a single vector, we can define more than one different vector by differentiation. The derivative of **A** depends on **A** *and* the reference frame from which its change is observed.

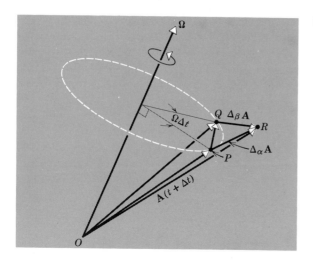

Figure 14-2

If reference frame β is translating, but not rotating, relative to frame α, the β-observed derivative of a vector will be equal to the α-observed derivative of the vector. This follows from the fact that the vector is defined only in terms of the relative positions of its end points in each reference frame, without regard for the position of the head of the vector. (See Section 3-1.)

Whenever there is more than one reference frame under consideration, we will denote the α-observed derivative of a vector as

$$\frac{d_\alpha \mathbf{A}}{dt} = \overset{\alpha}{\mathbf{A}} \qquad\qquad [9\text{-}2]^*$$

When only one reference frame can be inferred from the context of the discussion, the derivative of the vector $\mathbf{A}$ in that reference frame will be denoted as

$$\frac{d\mathbf{A}}{dt} = \dot{\mathbf{A}} \qquad\qquad [9\text{-}3]$$

In Newtonian dynamics, we will use this last notation to indicate derivatives observed from an inertial reference frame. The derivative of a scalar p is independent of the reference frame, and will be denoted as

$$\frac{dp}{dt} = \dot{p} \qquad\qquad [9\text{-}4]$$

* Brackets signify an equation that is repeated from an earlier development.

Now consider the situation in which a reference frame β rotates with angular velocity $\mathbf{\Omega}$ relative to reference frame α. The relationship between the α-observed derivative of a vector $\mathbf{A}(t)$ and the β-observed derivative of the vector may be inferred from Figure 14-2. This figure shows the relative positions, at time $t + \Delta t$, of the vector $\mathbf{A}$ and two vectors that were at time t coincident with $\mathbf{A}$, one being fixed in each reference frame. That is, at time t, points P, Q, and R were coincident; point P is fixed in reference frame α, point Q is fixed in reference frame β, and point R remains at the head of the vector $\mathbf{A}$. The α-observed change in $\mathbf{A}$ is the vector from point P to point R and is denoted by $\Delta_\alpha \mathbf{A}$; the β-observed change in $\mathbf{A}$ is the vector from point Q to point R and is denoted by $\Delta_\beta \mathbf{A}$. The difference is the vector from point P to point Q, and is approximately equal to $\mathbf{\Omega} \times \mathbf{A}\Delta t$. Thus,

$$\Delta_\alpha \mathbf{A} \approx \Delta_\beta \mathbf{A} + \mathbf{\Omega} \times \mathbf{A}\Delta t$$

As $\Delta t \to 0$, the approximation becomes arbitrarily accurate, and we have

$$\frac{d_\alpha \mathbf{A}}{dt} = \frac{d_\beta \mathbf{A}}{dt} + \mathbf{\Omega} \times \mathbf{A}$$

or

$$\boxed{\overset{\alpha}{\mathbf{A}} = \overset{\beta}{\mathbf{A}} + \mathbf{\Omega} \times \mathbf{A}} \qquad (14\text{-}1)$$

This relationship is fundamental to all analysis involving moving reference frames. In the general study of kinematics in the next section it will be applied to position and velocity vectors; in the study of rigid body kinetics (Chapter 16) it will be applied to angular momentum vectors.

Problems

14-1   A vector has components given by

$$\mathbf{A} = \cos \omega t \mathbf{u}_x + \sin \omega t \mathbf{u}_y + bt \mathbf{u}_z$$

where the x, y, and z axes are fixed in a reference frame β, which is rotating relative to a reference frame α with angular velocity

$$_\alpha \mathbf{\Omega}_\beta = 3\omega \mathbf{u}_z$$

Evaluate $\overset{\beta}{\mathbf{A}}$, $\overset{\alpha}{\underset{\beta}{\mathbf{A}}}$,$\overset{\alpha}{\mathbf{A}}$, and $\overset{\beta}{\underset{\alpha}{\mathbf{A}}}$

14-2   Repeat Problem 14-1, except that now the x, y, and z axes are fixed to the reference frame α.

14-3   A reference frame β with the coordinate axes x, y, and z attached is rotating relative to reference frame α with angular velocity $\mathbf{\Omega}$. Thus, the α-observed rates of change of the unit base vectors, $\mathbf{u}_x$, $\mathbf{u}_y$, and $\mathbf{u}_z$ are

$$\overset{\alpha}{\dot{\mathbf{u}}}_x = \mathbf{\Omega} \times \mathbf{u}_x \qquad \overset{\alpha}{\dot{\mathbf{u}}}_y = \mathbf{\Omega} \times \mathbf{u}_y \qquad \overset{\alpha}{\dot{\mathbf{u}}}_z = \mathbf{\Omega} \times \mathbf{u}_z$$

while their β-observed rates of change are zero. Evaluate the α-observed derivative of each member of the equation

$$\mathbf{A} = A_x\mathbf{u}_x + A_y\mathbf{u}_y + A_z\mathbf{u}_z$$

and deduce Equation 14-1 from the result.

14-4   Verify that $\overset{\beta}{\dot{\mathbf{A}}}\cdot\mathbf{B} + \mathbf{A}\cdot\overset{\beta}{\dot{\mathbf{B}}} = \overset{\alpha}{\dot{\mathbf{A}}}\cdot\mathbf{B} + \mathbf{A}\cdot\overset{\alpha}{\dot{\mathbf{B}}}$. Hence the rotation rate of the reference frame has no influence on the drivative of the scalar $\mathbf{A}\cdot\mathbf{B}$.

14-5   Verify that

$$\overset{\alpha}{\dot{\mathbf{A}}} \times \mathbf{B} + \overset{\alpha}{\dot{\mathbf{B}}} \times \mathbf{A} = \mathbf{A} \times \overset{\beta}{\dot{\mathbf{B}}} + \overset{\beta}{\dot{\mathbf{A}}} \times \mathbf{B} + {}_\alpha\mathbf{\Omega}_\beta \times (\mathbf{A} \times \mathbf{B})$$

Hence Equation 14-1 is consistent with Equation 9-8.

14-2

PARTICLE KINEMATICS. In defining the velocity of a point as the derivative of a position vector, two items are associated with the reference frame. The position vector must locate the moving point from some point fixed in that reference frame, *and* the derivative must be observed in the same reference frame. Thus, we define the α-observed velocity of point P as

$$_\alpha\mathbf{v}_P = \overset{\alpha}{\dot{\mathbf{r}}}_{P/A} \tag{14-2}$$

where A is any point fixed in the reference frame α.

Example

The airplane β in Figure 14-3a is traveling at 900 km/h in a banking turn of radius 3.7 km. A second airplane P is traveling at 1400 km/h parallel (at the instant shown) to the first. What is the velocity of airplane P observed from airplane β?

The second airplane may be treated as a point for purposes of the question here; however, the airplane β must be considered as a three-dimensional rigid body, or reference frame, because its changing orientation affects the velocities of points observed from it. This reference frame β could be thought of as a turntable with a radius of 3.7 km. Let us call α the reference frame attached to the ground, and define points A and B to be fixed in reference frames α and β, respectively, as shown in Figure 14-3b.

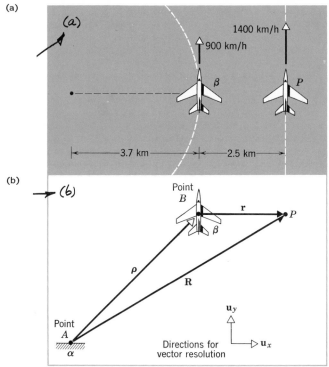

Figure 14-3

The position relationship

$$\mathbf{R} = \boldsymbol{\rho} + \mathbf{r}$$

may be differentiated to yield

$$\overset{\alpha}{\mathbf{R}} = \overset{\alpha}{\boldsymbol{\rho}} + \overset{\alpha}{\mathbf{r}} \tag{a}$$

The α-observed derivatives of $\mathbf{R}$ and $\boldsymbol{\rho}$ are readily determined from the given information, as

$$\overset{\alpha}{\mathbf{R}} = {}_{\alpha}\mathbf{v}_P = 1400 \text{ km/h } \mathbf{u}_y \tag{b}$$

$$\overset{\alpha}{\boldsymbol{\rho}} = {}_{\alpha}\mathbf{v}_B = 900 \text{ km/h } \mathbf{u}_y \tag{c}$$

However, the velocity we seek here is the *β-observed* derivative of $\mathbf{r}$, not $\overset{\alpha}{\mathbf{r}}$. Applying Equation 14-1 to $\mathbf{r}$, we obtain

$$\overset{\alpha}{\mathbf{r}} = \overset{\beta}{\mathbf{r}} + \left(\frac{900 \text{ km/h}}{3.7 \text{ km}} \mathbf{u}_z\right) \times (2.5 \text{ km } \mathbf{u}_x) \tag{d}$$

Substitution of (b), (c), and (d) into (a) yields

$$1400 \text{ km/h } \mathbf{u}_y = 900 \text{ km/h } \mathbf{u}_y + \overset{\beta}{\dot{\mathbf{r}}} + \frac{(900)(2.5)}{3.7} \text{ km/h } \mathbf{u}_y$$

Solving for the desired velocity, we obtain

$$_{\beta}\mathbf{v}_P = \overset{\beta}{\dot{\mathbf{r}}} = -108 \text{ km/h } \mathbf{u}_y$$

Thus the airplane P appears, from airplane β, to be falling behind at the rate of 108 km/h.

The acceleration of a point P in a reference frame α is defined as the α-observed derivative of the α-observed velocity of P:

$$_{\alpha}\mathbf{a}_P = {}_{\alpha}\overset{\alpha}{\dot{\mathbf{v}}}_P = \overset{\alpha\alpha}{\ddot{\mathbf{r}}}_{P/A} \tag{14-3}$$

Example

Determine the earth-observed velocity and acceleration of a satellite that, from an inertial reference frame, is travelling at 7617 m/s in a circular polar orbit of radius 6880 km, as shown in Figure 14-4a. Here we refer to an inertial reference frame as one in which the earth appears to make one revolution per day about a fixed polar axis; that is, this reference frame revolves about the sun once each year.

As indicated in Figure 14-4b, let us denote the inertial reference frame as α and an earth-fixed reference frame as β; let point O be fixed to the center of the earth (fixed to both α and β) and point P be fixed to the satellite; let the

Figure 14-4

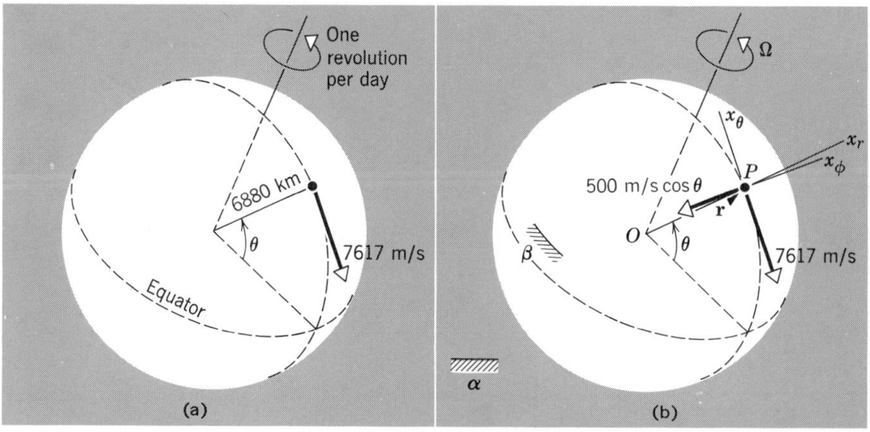

(a)

(b)

local upward, eastward, and northward directions be denoted as x_r, x_ϕ, and x_θ, respectively. Then the earth-observed velocity is defined as

$$\beta \mathbf{v}_P = \overset{\beta}{\mathbf{r}}$$

But, from Equation 14-1,

$$\overset{\beta}{\mathbf{r}} = \overset{\alpha}{\mathbf{r}} - \mathbf{\Omega} \times \mathbf{r}$$

in which the angular velocity of β relative to α is

$$\mathbf{\Omega} = \frac{2\pi \text{ rad}}{24 \text{ h}} \, (\cos \theta \, \mathbf{u}_\theta + \sin \theta \, \mathbf{u}_r)$$

$$= 7.272 \times 10^{-5} \text{ rad/s } (\cos \theta \, \mathbf{u}_\theta + \sin \theta \, \mathbf{u}_r)$$

Furthermore,

$$\overset{\alpha}{\mathbf{r}} = {}_\alpha \mathbf{v}_P = -7617 \text{ m/s } \mathbf{u}_\theta$$

and

$$\mathbf{r} = 6.88 \times 10^6 \text{ m } \mathbf{u}_r$$

Substitution of these values above results in

$$\beta \mathbf{v}_P = \overset{\alpha}{\mathbf{r}} - \mathbf{\Omega} \times \mathbf{r}$$

$$= -7617 \text{ m/s } \mathbf{u}_\theta - 500 \text{ m/s } \cos \theta \, \mathbf{u}_\phi$$

The earth-observed acceleration is defined as

$$\beta \mathbf{a}_P = {}_\beta \overset{\beta}{\mathbf{v}}_P$$

Using Equation 14-1 again, and substituting from the velocity analysis, we can rewrite this as

$$\overset{\beta}{}_\beta \mathbf{v}_P = {}_\beta \overset{\alpha}{\mathbf{v}}_P - \mathbf{\Omega} \times {}_\beta \mathbf{v}_P$$

$$= \frac{d_\alpha}{dt} (\overset{\alpha}{\mathbf{r}} - \mathbf{\Omega} \times \mathbf{r}) - \mathbf{\Omega} \times {}_\beta \mathbf{v}_P$$

$$= {}_\alpha \mathbf{a}_P - \mathbf{\Omega} \times \overset{\alpha}{\mathbf{r}} - \mathbf{\Omega} \times {}_\beta \mathbf{v}_P$$

Now the inertial-observed acceleration is

$$_\alpha \mathbf{a}_P = - \frac{(7617 \text{ m/s})^2}{(6.88 \times 10^6 \text{m})} \, \mathbf{u}_r$$

$$= - 8.43 \text{ m/s}^2 \, \mathbf{u}_r$$

The remaining terms in the above may be evaluated from the results of the velocity analysis:

$$-\mathbf{\Omega} \times \overset{\alpha}{\dot{\mathbf{r}}} - \mathbf{\Omega} \times {}_\beta\mathbf{v}_P = -\mathbf{\Omega} \times ({}_\alpha\mathbf{v}_P + {}_\beta\mathbf{v}_P)$$

$$= -(7.272 \times 10^{-5} \text{ rad/s}) (\cos \theta \, \mathbf{u}_\theta + \sin \theta \, \mathbf{u}_r)$$
$$\times (-15\,234 \text{ m/s } \mathbf{u}_\theta - 500 \text{ m/s } \cos \theta \, \mathbf{u}_\phi)$$

$$= -(0.04 \cos \theta \sin \theta \, \mathbf{u}_\theta$$
$$- 1.11 \sin \theta \, \mathbf{u}_\phi - 0.04 \cos^2 \theta \, \mathbf{u}_r) \text{m/s}^2$$

Substitution of these values into the above yields the value of the earth-observed acceleration:

$${}_\beta\mathbf{a}_P = [0.04 \cos \theta \sin \theta \, \mathbf{u}_\theta - 1.11 \sin \theta \, \mathbf{u}_\phi - (8.43 + 0.04 \cos^2 \theta)\mathbf{u}_r] \text{m/s}^2$$

The somewhat-strange westward component $-1.11 \sin \theta \text{ m/s}^2 \, \mathbf{u}_\phi$ is evident in the curvature of the path in Figure 14-1.

General Relationships for Velocity and Acceleration.   A general description of velocities and accelerations from moving reference frames can be obtained by application of Equation 14-1 as in the preceding examples. The resulting formulas may be used to reduce analysis of this type to straightforward substitutions.

Consider the general motion of a point P as observed from two different reference frames, as depicted in Figure 14-5. Points A and B are fixed in reference frames α and β, respectively. The motion of β relative to α is given by a

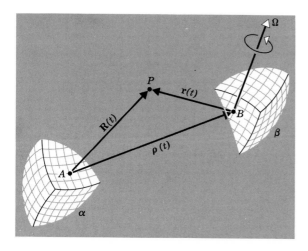

Figure 14-5

simultaneous variation of $\rho(t)$ and a rotation of β about an axis through B at the angular velocity $\mathbf{\Omega}$. We will see in Section 14-4 that this comprises the most general motion β can have.

Differentiation of the position relationship

$$\mathbf{R} = \boldsymbol{\rho} + \mathbf{r}$$

results in

$$\overset{\alpha}{\mathbf{R}} = \overset{\alpha}{\boldsymbol{\rho}} + \overset{\alpha}{\mathbf{r}}$$

Two of the terms may be readily recognized as α-observed velocities:

$$\overset{\alpha}{\mathbf{R}} = {}_{\alpha}\mathbf{v}_P \qquad \overset{\alpha}{\boldsymbol{\rho}} = {}_{\alpha}\mathbf{v}_B$$

However, the last term, the α-observed rate of change of the β-observed position, is neither easily understood nor easily evaluated as it stands. But an application of Equation 14-1 yields two components that are readily handled:

$$\overset{\alpha}{\mathbf{r}} = \overset{\beta}{\mathbf{r}} + \mathbf{\Omega} \times \mathbf{r} = {}_{\beta}\mathbf{v}_P + \mathbf{\Omega} \times \mathbf{r}$$

Thus, we have the general velocity relationship

$$\boxed{{}_{\alpha}\mathbf{v}_P = {}_{\beta}\mathbf{v}_P + {}_{\alpha}\mathbf{v}_B + \mathbf{\Omega} \times \mathbf{r}} \qquad (14\text{-}4)$$

Differentiation of Equation 14-4 gives the acceleration relationship

$$\overset{\alpha}{{}_{\alpha}\mathbf{v}_P} = \overset{\alpha}{{}_{\beta}\mathbf{v}_P} + \overset{\alpha}{{}_{\alpha}\mathbf{v}_B} + \mathbf{\Omega} \times \overset{\alpha}{\mathbf{r}} + \overset{\alpha}{\mathbf{\Omega}} \times \mathbf{r}$$

Two of the terms are readily recognized as α-observed accelerations:

$$\overset{\alpha}{{}_{\alpha}\mathbf{v}_P} = {}_{\alpha}\mathbf{a}_P \qquad \overset{\alpha}{{}_{\alpha}\mathbf{v}_B} = {}_{\alpha}\mathbf{a}_B$$

As in the velocity analysis, application of Equation 14-1 to the remaining terms yields components that have meaning and are readily handled:

$$\overset{\alpha}{{}_{\beta}\mathbf{v}_P} = \overset{\beta}{{}_{\beta}\mathbf{v}_P} + \mathbf{\Omega} \times {}_{\beta}\mathbf{v}_P = {}_{\beta}\mathbf{a}_P + \mathbf{\Omega} \times {}_{\beta}\mathbf{v}_P$$

$$\mathbf{\Omega} \times \overset{\alpha}{\mathbf{r}} = \mathbf{\Omega} \times (\overset{\beta}{\mathbf{r}} + \mathbf{\Omega} \times \mathbf{r}) = \mathbf{\Omega} \times {}_{\beta}\mathbf{v}_P + \mathbf{\Omega} \times (\mathbf{\Omega} \times \mathbf{r})$$

Substitution into the above yields the general acceleration relationship

$$\boxed{{}_{\alpha}\mathbf{a}_P = {}_{\beta}\mathbf{a}_P + {}_{\alpha}\mathbf{a}_B + \overset{\alpha}{\mathbf{\Omega}} \times \mathbf{r} + 2\mathbf{\Omega} \times {}_{\beta}\mathbf{v}_P + \mathbf{\Omega} \times (\mathbf{\Omega} \times \mathbf{r})} \quad (14\text{-}5)$$

Example

Determine the velocity and acceleration of a point moving at speed $v(t)$ in a plane circular path of radius ρ.

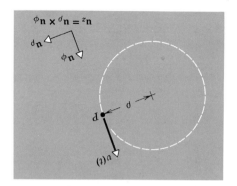

Let β rotate about the center of the circle at a rate such that the point is fixed in this reference frame, and let the points A and B be at the center of the circle. Then the vectors in the velocity equation (14-4) have the values

$$_\beta\mathbf{v}_P = {_\alpha}\mathbf{v}_B = \mathbf{0}$$

$$\mathbf{\Omega} = \frac{v}{\rho}\,\mathbf{u}_z$$

$$\mathbf{r} = \rho\mathbf{u}_\rho$$

so that the velocity observed from the fixed reference frame is

$$_\alpha\mathbf{v}_P = \mathbf{0} + \mathbf{0} + \frac{v}{\rho}\,\mathbf{u}_z \times \rho\mathbf{u}_\rho$$

$$= v\,\mathbf{u}_\phi$$

The vectors in the acceleration equation (14-5) have the values

$$_\beta\mathbf{a}_P = {_\alpha}\mathbf{a}_B = {_\beta}\mathbf{v}_P = \mathbf{0}$$

$$\mathbf{\Omega} = \frac{v}{\rho}\,\mathbf{u}_z$$

$$\overset{\alpha}{\mathbf{\Omega}} = \frac{\dot{v}}{\rho}\,\mathbf{u}_z$$

The acceleration observed from the fixed reference frame is then

$$_\alpha\mathbf{a}_P = \mathbf{0} + \mathbf{0} + \frac{\dot{v}}{\rho}\,\mathbf{u}_z \times \rho\mathbf{u}_\rho + \mathbf{0} + \frac{v}{\rho}\,\mathbf{u}_z \times \left(\frac{v}{\rho}\,\mathbf{u}_z \times \rho\mathbf{u}_\rho\right)$$

$$= \dot{v}\,\mathbf{u}_\phi - \frac{v^2}{\rho}\,\mathbf{u}_\rho$$

The term $\mathbf{\Omega} \times (\mathbf{\Omega} \times \mathbf{r})$, having the value $-\,(v^2/\rho)\mathbf{u}_\rho$ in this example, is called the centripetal component of acceleration. As we saw in Chapter 10, it arises through the changing direction of the tangential component of velocity.

Example

A particle P is constrained to move along a straight radial line relative to the rotating disc, as shown in Figure 14-6. The disc rotates at a constant rate ω about an axis perpendicular to the groove. Determine the velocity and acceleration of P as observed from the reference frame in which the disc is rotating.

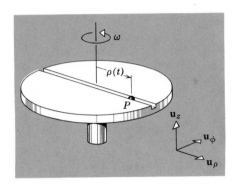

Figure 14-6

Let α be the fixed reference frame and β be a reference frame attached to the disc. Also, let points A and B coincide at the center of the disc. Then the vectors in Equation 14-4 have the values

$$_\beta\mathbf{v}_P = \dot{\rho}\mathbf{u}_\rho \qquad _\alpha\mathbf{v}_B = \mathbf{0}$$

$$\mathbf{\Omega} = \omega\mathbf{u}_z \qquad \mathbf{r} = \rho\mathbf{u}_\rho$$

so that the velocity observed from the fixed reference frame is

$$_\alpha\mathbf{v}_P = \dot{\rho}\mathbf{u}_\rho + \rho\omega\mathbf{u}_\phi$$

The vectors in Equation 14-5 have the values

$$_\beta\mathbf{a}_P = \ddot{\rho}\mathbf{u}_\rho \qquad _\alpha\mathbf{a}_B = \overset{\alpha}{\dot{\mathbf{\Omega}}} = \mathbf{0}$$

$$\mathbf{\Omega} = \omega\mathbf{u}_z \qquad _\beta\mathbf{v}_P = \dot{\rho}\mathbf{u}_\rho$$

$$\mathbf{r} = \rho\mathbf{u}_\rho$$

The acceleration observed from the fixed reference frame is then determined by substitution into Equation 14-5:

$$_\alpha\mathbf{a}_P = (\ddot{\rho} - \rho\omega^2)\mathbf{u}_\rho + 2\omega\dot{\rho}\mathbf{u}_\phi$$

Coriolis component of acceleration. The term $2\omega\dot{\rho}\mathbf{u}_\phi$ in the last example seems curious at first. It can be best understood as the sum of two effects. Figure 14-7 shows the position of P and the radial and tangential components of its velocity (relative to the fixed reference frame) at time t and at time $t + \Delta t$. First,

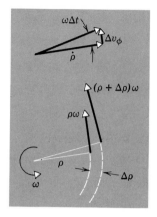

observe that the radial component undergoes a change in the tangential direction, due to the changing direction of the groove:

$$\Delta v_\phi = \dot{\rho}\omega\Delta t$$

Second, due to the change in radial position of P, the tangential speed $\rho\omega$ undergoes a change

$$\Delta v_\phi = (\rho + \Delta\rho)\omega - \rho\omega$$

These two contributions combine to produce the component $2\omega\dot{\rho}\mathbf{u}_\phi$.

The term $2\,\boldsymbol{\Omega}\times {}_\beta\mathbf{v}_P$ in the acceleration equation (14-5) is called the *Coriolis* component of acceleration, after the French physicist Gaspard de Coriolis (1792–1843), who studied the motions of bodies on spinning surfaces. The two terms that combine in the derivation to form the Coriolis term in Equation 14-5 can be interpreted in the same way as the two effects contributing to the Coriolis component $2\,\omega\dot{\rho}\mathbf{u}_\phi$ in the above example.

Example

Determine the velocity and acceleration of airplane P as observed from airplane β, at the instant the airplanes are in the positions shown in Figure 14-3. Both airplanes are traveling at constant speed.

With the same choice of reference frames and points used earlier, the vectors in Equation 14-4 have the values

$$_\alpha\mathbf{v}_P = 1400 \text{ km/h } \mathbf{u}_y$$

$$_\alpha\mathbf{v}_B = 900 \text{ km/h } \mathbf{u}_y$$

$$\boldsymbol{\Omega} = \left(\frac{900 \text{ km/h}}{3.7 \text{ km}}\right)\mathbf{u}_z$$

$$\mathbf{r} = 2.5 \text{ km } \mathbf{u}_x$$

Substitution into Equation 14-4 gives the value of $_\beta \mathbf{v}_P$ determined earlier. The vectors in Equation 14-5 have the values

$$_\alpha \mathbf{a}_P = \mathbf{0}$$

$$_\alpha \mathbf{a}_B = -\frac{(900 \text{ km/h})^2}{3.7 \text{ km}} \mathbf{u}_x = -16.89 \text{ m/s}^2 \ \mathbf{u}_x$$

$$^\alpha \boldsymbol{\Omega} = \mathbf{0}$$

$$\boldsymbol{\Omega} = 0.0676 \text{ rad/s } \mathbf{u}_z$$

$$_\beta \mathbf{v}_P = -30.03 \text{ m/s } \mathbf{u}_y$$

$$\mathbf{r} = 2500 \text{ m } \mathbf{u}_x$$

Substitution into Equation 14-5 leads to the value

$$_\beta \mathbf{a}_P = 24.2 \text{ m/s}^2 \ \mathbf{u}_x$$

Example

Determine the earth-observed velocity and acceleration of the satellite shown in Figure 14-4.

With the same choice of reference frames used earlier, and with points A and B both at the center of the earth, the vectors in Equation 14-4 have the values

$$_\alpha \mathbf{v}_P = -7617 \text{ m/s } \mathbf{u}_\theta$$

$$_\alpha \mathbf{v}_B = \mathbf{0}$$

$$\boldsymbol{\Omega} = 7.272 \times 10^{-5} \text{ rad/s } (\cos \theta \ \mathbf{u}_\theta + \sin \theta \ \mathbf{u}_r)$$

$$\mathbf{r} = 6.880 \times 10^6 \text{ m } \mathbf{u}_r$$

Substitution into Equation 14-4 leads to the value of the earth-observed velocity obtained earlier. Vectors in addition to those just evaluated, which appear in Equation 14-5, are

$$_\alpha \mathbf{a}_P = -\frac{(7617 \text{ m/s})^2}{6880 \text{ km}} \mathbf{u}_r = -8.433 \text{ m/s}^2 \ \mathbf{u}_r$$

$$_\alpha \mathbf{a}_B = {}^\alpha \boldsymbol{\Omega} = \mathbf{0}$$

Substitution into Equation 14-5 then leads to the value of the earth-observed acceleration obtained earlier. The westward component of the earth-observed acceleration, $-1.11 \sin \theta \text{ m/s}^2 \ \mathbf{u}_\phi$, arises from the Coriolis term in Equation 14.5.

Problems

14-6 A disc is spinning in a horizontal plane, and a small particle travels at constant speed in a straight, horizontal line over the spin center. A photographic trace of the particle is made on the disc. Sketch the trace on the disc, indicating the direction the disc was spinning. Indicate the direction of the acceleration of the particle as observed from the disc.

14-7 A saucer-shaped UFO was sighted as its center moved at constant speed v in a horizontal circular path of radius R. It was simultaneously spinning about a vertical axis through its center at the rate ω rad/s.

(a) Graphically locate the instantaneous center of the UFO, and from this determine the ground-observed velocity of a point on the rim of the saucer.

(b) Determine the ground-observed acceleration of this point.

14-8 An automobile accelerates from rest on a slippery surface at 1 m/s², by spinning its wheels at a constant 25 rad/s. The radius of the wheels is 380 mm. Determine the velocity and acceleration that the point on a wheel in contact with the surface will have, 4 s after starting.

14-9 Using a reference frame β attached to the unit vectors $\mathbf{u}_\rho$ and $\mathbf{u}_\phi$ of the cylindrical coordinates, determine the α-observed velocity and acceleration of the particle P, using Equations 14-4 and 14-5.

14-10 In Section 14-4, it will be shown that the angular velocity of the reference frame attached to the triad $\mathbf{u}_r - \mathbf{u}_\chi - \mathbf{u}_\phi$ is equal to the vector sum of the components shown with magnitudes $\dot\phi$ and $\dot\chi$. Use (14-4) and (14-5) to determine the α-observed velocity and acceleration of the point P, in terms of r, ϕ, and χ and their derivatives.

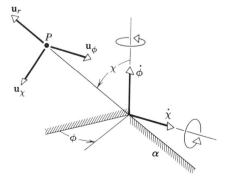

14-11   Compare the results of Problem 14-10 with Equations 10-8.

14-12   Under what circumstances can the velocity and acceleration differences,

$$\mathbf{v}_{P/B} \equiv \mathbf{v}_P - \mathbf{v}_B$$

$$\mathbf{a}_{P/B} \equiv \mathbf{a}_P - \mathbf{a}_B$$

be properly termed *relative* velocity and acceleration? Explain in terms of Equations 14-4 and 14-5.

14-13   Referring to the intrinsic coordinates,* let us examine the way in which the $\mathbf{u}_t - \mathbf{u}_n - \mathbf{u}_b$ triad rotates. Observe that if we call its angular velocity $\mathbf{\Omega}$, we have

$$\dot{\mathbf{u}}_t = \mathbf{\Omega} \times \mathbf{u}_t \qquad \dot{\mathbf{u}}_n = \mathbf{\Omega} \times \mathbf{u}_n \qquad \dot{\mathbf{u}}_b = \mathbf{\Omega} \times \mathbf{u}_b$$

(a) Explain why the angular velocity must lie in the $\mathbf{u}_b - \mathbf{u}_t$ plane.

(b) Show that the binormal component of $\mathbf{\Omega}$ has the value

$$\Omega_b = v\kappa$$

(c) The triad can also rotate about an axis in the direction of $\mathbf{u}_t$. We define the *torsion* τ of the space curve as a measure of this component of rotation in the same way that the curvature measures the b component of rotation:

$$\Omega_t = v\tau$$

Show that

$$\frac{d\mathbf{u}_t}{ds} = \kappa\mathbf{u}_n$$

$$\frac{d\mathbf{u}_n}{ds} = -\kappa\mathbf{u}_t + \tau\mathbf{u}_b$$

$$\frac{d\mathbf{u}_b}{ds} = -\tau\mathbf{u}_n$$

14-14   From the previous problem we have the angular velocity of the $\mathbf{u}_t - \mathbf{u}_n - \mathbf{u}_b$ triad, in terms of the curvature κ and the torsion τ:

$$\mathbf{\Omega} = v(\kappa\mathbf{u}_b + \tau\mathbf{u}_t)$$

Derive the intrinsic components of acceleration by applying Equation 14-1 to the vector $\mathbf{v}$.

14-15   The time rate of change of acceleration is sometimes called the "jerk." Show that it has the intrinsic components

$$\mathbf{J} = (\ddot{v} - \kappa^2 v^3)\mathbf{u}_t + (\dot{\kappa}v^2 + 3\kappa v\dot{v})\mathbf{u}_n + (\kappa\tau v^3)\mathbf{u}_b$$

*Charles Smith, *Dynamics*, Wiley, 1976, pp. 21–23.

14-16   We have pointed out that the translational motion of a reference frame has no influence on the derivative of a vector. Yet the velocity of a particle, which is a derivative of a position vector, certainly depends on the translational motion of the observer. Resolve this apparent paradox.

14-17   In the indexing mechanism shown the indexing finger BP slides back and forth on the fixed pin A as arm OB rotates about O at 200 rpm. Determine the velocity and acceleration of end P for θ_{max} and θ_{min}.

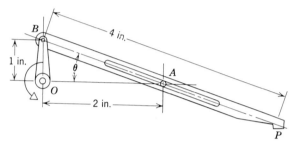

14-18   The lower link drives the upper link through the slider connection. In the position shown (all dimensions given), determine the angular velocity of the upper link in terms of that of the lower link.

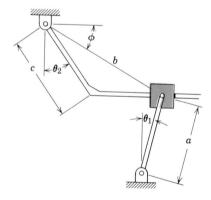

14-19   Evaluate the velocity and acceleration of the slider pin in the mechanism of the previous problem, in terms of the angular velocity $\omega_2(t)$ of the upper link and the given dimensions. *Suggestion*: A reference frame attached to the upper link will be helpful.

14-20   Show that we can replace $\overset{\alpha}{\Omega}$ with $\overset{\beta}{\Omega}$ in Equation 14-5.

14-21   Airplane A is making a banking maneuver at constant speed v, while airplane B is making a loop maneuver at constant speed $2v/3$. Evaluate the velocity and acceleration of B, as observed by one riding in A, at the instant the airplanes are in the positions shown.

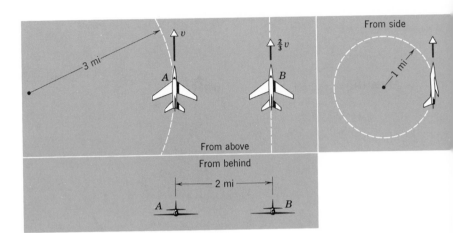

14-22

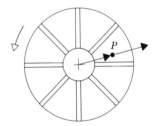

Fluid particle P moves with a constant radial velocity v relative to the centrifugal pump impeller, which turns with a constant angular velocity. Determine the velocity and acceleration of P.

14-23 The time rate of change of acceleration is sometimes called the "jerk" $\mathbf{J}$. Derive the relationship between the α-observed "jerk" $_\alpha\mathbf{J}_P$ of a moving point P, and its β-observed "jerk" $_\beta\mathbf{J}_P$.

14-24 The cylinder rotates about a fixed point B at a constant ω rad/s. The connecting rod AC rotates about the fixed point C as the piston slides back and forth in the cylinder. What will be the acceleration of point A when $\theta = 30°$?

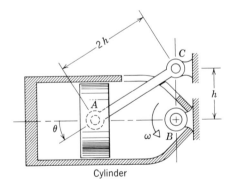

Cylinder

14-25   The disk shown rotates at angular speed $\Omega = \Omega_0 - \alpha t$ as the pin moves in the slot at speed $v = Bt^2$. Find the acceleration of the pin at the instant shown corresponding to time $t = t_1$.

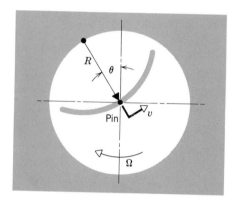

14-26   The pin A moves in the slot with a constant velocity v relative to the slot in the direction shown. The disk turns clockwise about its center at an angular velocity of $\omega = \omega_0 + \alpha t$. Determine the velocity and acceleration of the pin at the time the pin is in the position shown.

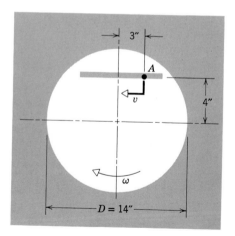

14-27   The pin moves along the slot with constant speed v relative to the slot as the disc rotates with angular velocity $\Omega = \alpha t$. Find the acceleration of the pin at the instant shown, corresponding to time $t = t_1$.

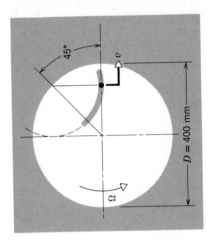

14-28 The disc rotates at a constant Ω rad/s. The slider P is constrained to move along the horizontal guide. Determine the velocity and acceleration of P.

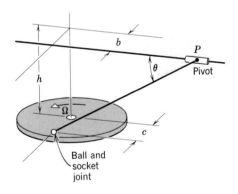

14-29 The barrel of a gun is swinging about its base at the constant rate of 4 rad/s. A projectile follows a curved path as it traverses the horizontal barrel. Determine the component of the projectile's acceleration normal to the barrel's axis just before the projectile emerges at its muzzle velocity of 6000 ft/s.

14-30   The airplane is banking as shown. Evaluate the ground-observed velocity and acceleration of the tip of the propeller at the instant it points toward the center of the turn.

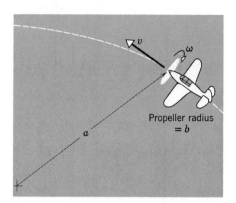

14-31   Airplane C is banking in a horizontal plane at constant speed v_C while airplane D is "looping" at constant speed v_D. At the instant shown what are the velocity and acceleration of D as observed by C? What is the velocity of C as observed by D?

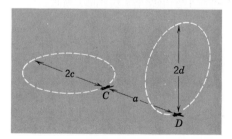

14-32   The destroyer D is making a turn while watching the submarine S, which is diving uniformly in a straight line. It is also rolling with sinusoidal motion as indicated in the rear view. Evaluate the destroyer-observed velocity and acceleration of S.

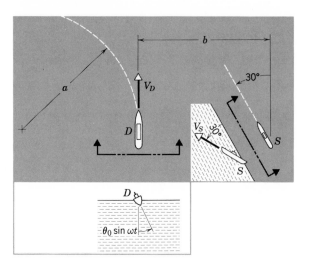

14-33 The destroyer D is making a turn while watching the submarine S, which is diving uniformly in a straight line. It is also rolling with approximately sinusoidal motion, as indicated in the rear view. At the instant the destroyer is rolled hard over to port, and the vessels are in the position shown in the top view, what velocity and acceleration of the submarine does an observer riding on the destroyer see?

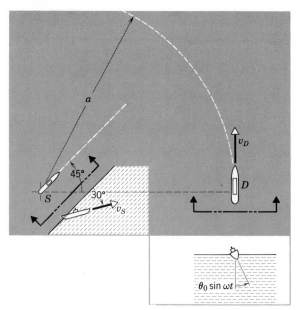

14-3

PARTICLE KINETICS IN ROTATING REFERENCE FRAMES. In addition to simplifying certain problems in kinematics, the velocity and acceleration relationships (14-4) and (14-5) are useful for predicting motions as they will be viewed from such moving reference frames as vehicles, or the rotating earth. Before becoming involved in detailed analysis, let us visualize, in a simple way, some of the effects that can arise from motion of a reference frame.

Suppose we toss an object into the air inside a rocket ship that is in deep space and accelerating. Since the acceleration of the object in an inertial reference frame will be zero, it will appear to us to be accelerating toward the rear of the ship, just as objects we toss up near the surface of the earth accelerate downward. In fact, there is no mechanical means by which we could distinguish between the effect of our accelerating reference frame and a gravitational attraction of like magnitude. To everyone and everything aboard, "down" would be toward the rear of the rokcet.

Rotation of a reference frame will produce more complicated effects. For example, the satellite shown in Figure 14-1 appears from the rotating earth to have a westward component of acceleration, while from an inertial reference frame its only acceleration is that directed toward the center of the earth.

The analysis of such motions is readily accomplished with Newton's second law and the kinematic relationships that we studied in the previous section. With inertial reference frames denoted by the subscript 0 and inertial-observed derivatives by dots, Newton's second law can be written as

$$\mathbf{f} = m[{}_{\beta}\mathbf{a}_P + {}_{0}\mathbf{a}_B + \dot{\boldsymbol{\Omega}} \times \mathbf{r} + 2\boldsymbol{\Omega} \times {}_{\beta}\mathbf{v}_P + \boldsymbol{\Omega} \times (\boldsymbol{\Omega} \times \mathbf{r})] \qquad (14\text{-}6)$$

Example

Write the differential equation that governs the angle $\phi(t)$ as the pendulum oscillates inside the accelerating vehicle shown in Figure 14-7.

With the truck as the moving reference frame β, we have

$$_{\beta}\mathbf{a}_P = l\ddot{\phi}\mathbf{u}_\phi - l\dot{\phi}^2\mathbf{u}_\rho$$

$$_{0}\mathbf{a}_B = -a_B\mathbf{u}_x \qquad \boldsymbol{\Omega} = 0$$

so that the inertial-observed acceleration of the particle is

$$_{0}\mathbf{a}_P = {}_{\beta}\mathbf{a}_P + {}_{0}\mathbf{a}_B$$

$$= l\ddot{\phi}\mathbf{u}_\phi - l\dot{\phi}^2\mathbf{u}_\rho - a_B\mathbf{u}_x$$

Figure 14-7

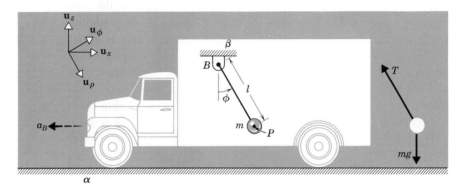

From the free-body diagram, we see that

$$\mathbf{f} = -T\mathbf{u}_\rho - f_g\mathbf{u}_z$$

Thus, Newton's second law states that

$$-T\mathbf{u}_\rho - f_g\mathbf{u}_z = m(l\ddot{\phi}\mathbf{u}_\phi - l\dot{\phi}^2\mathbf{u}_\rho - a_B\mathbf{u}_x)$$

In order to express this in terms of a common set of directions, we substitute the unit vector resolutions

$$\mathbf{u}_x = \mathbf{u}_\phi \cos\phi + \mathbf{u}_\rho \sin\phi$$

$$\mathbf{u}_z = \mathbf{u}_\phi \sin\phi - \mathbf{u}_\rho \cos\phi$$

with the result

$$(mg\cos\phi - T)\mathbf{u}_\rho - mg\sin\phi\,\mathbf{u}_\phi = m[-(a_B\sin\phi + l\dot{\phi}^2)\mathbf{u}_\rho + (l\ddot{\phi} - a_B\cos\phi)\mathbf{u}_\phi]$$

This vector equation implies the two component equations

$$l\ddot{\phi} - a_B\cos\phi + g\sin\phi = 0$$

$$m(g\cos\phi + a_B\sin\phi + l\dot{\phi}^2) - T = 0$$

The first of these is the differential equation governing the angle ϕ, and the second gives the tension T in terms of the motion of the pendulum.

For a constant vehicle acceleration, an equilibrium position ϕ_0 is possible, satisfying

$$-a_B\cos\phi_0 + g\sin\phi_0 = 0$$

$$m(g\cos\phi_0 + a_B\sin\phi_0) - T = 0$$

or,

$$\phi_0 = \tan^{-1}\left(\frac{g}{a_B}\right)$$

$$T_0 = m\sqrt{g^2 + a_B{}^2}$$

To an observer inside the truck, mechanical behavior is as though a gravitational field were acting in the direction parallel to the pendulum in this equilibrium position.

Example

Examine the motion, viewed from the rotating earth β, of a particle P moving in the neighborhood of the latitude θ. Its position will be specified in the locally determined rectangular Cartesian coordinate system with axes x_r, x_ϕ, and x_θ fixed to the earth, as shown in Figure 14-8.

   With point B chosen at the origin of these coordinates, the several terms in Equation 14-6 will have the values

$$\mathbf{f} = -mg\mathbf{u}_r$$

$$_0\mathbf{a}_B = a\cos\theta\,\Omega^2(\sin\theta\,\mathbf{u}_\theta - \cos\theta\,\mathbf{u}_r)$$

$$_\beta\mathbf{a}_P = \ddot{x}_r\mathbf{u}_r + \ddot{x}_\phi\mathbf{u}_\phi + \ddot{x}_\theta\mathbf{u}_\theta$$

$$\dot{\boldsymbol{\Omega}} = 0$$

$$2\boldsymbol{\Omega} \times {}_\beta\mathbf{v}_P = 2\Omega(\cos\theta\,\mathbf{u}_\theta + \sin\theta\,\mathbf{u}_r) \times (\dot{x}_r\mathbf{u}_r + \dot{x}_\phi\mathbf{u}_\phi + \dot{x}_\theta\mathbf{u}_\theta)$$

$$= 2\Omega[-\dot{x}_\phi\cos\theta\,\mathbf{u}_r + (\dot{x}_r\cos\theta - \dot{x}_\theta\sin\theta)\mathbf{u}_\phi + \dot{x}_\phi\sin\theta\,\mathbf{u}_\theta]$$

$$|\boldsymbol{\Omega} \times (\boldsymbol{\Omega} \times \mathbf{r})| \ll |{}_0\mathbf{a}_B| \qquad \text{for } r \ll a\cos\theta$$

Figure 14-8

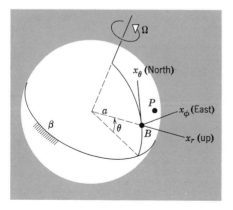

With the vectors thus resolved and $\mathbf{\Omega} \times (\mathbf{\Omega} \times \mathbf{r})$ neglected, the component equations implied by Equation 14-6 are

$$\ddot{x}_r - 2\Omega \cos \theta \; \dot{x}_\phi = -g + a\Omega^2 \cos^2 \theta \qquad \text{(a)}$$

$$\ddot{x}_\phi + 2\Omega(\cos \theta \; \dot{x}_r - \sin \theta \; \dot{x}_\theta) = 0 \qquad \text{(b)}$$

$$\ddot{x}_\theta + 2\Omega \sin \theta \; \dot{x}_\phi = -a\Omega^2 \cos \theta \sin \theta \qquad \text{(c)}$$

At the equator, the value of $a\Omega^2 \cos^2 \theta$ is approximately 0.3% of g, and so may reasonably be neglected in the first equation. Furthermore, in order for the term $2\Omega \cos \theta \; \dot{x}_\phi$ to be significant in comparison with $a\Omega^2 \cos \theta$ (which, we have just pointed out, is negligible) would require that $\dot{x}_\phi$ be of the order of $\frac{1}{2}a\Omega \simeq$ 200 m/s. Neglecting this term as well, we have

$$\ddot{x}_r = -g$$

from which

$$\dot{x}_r = v_0 - gt$$

$$x_r = v_0 t - \tfrac{1}{2}gt^2$$

Now let us predict the lateral displacement of an object hurled straight upward, that is,

$$\dot{x}_r(0) = v_0 \qquad \dot{x}_\phi(0) = \dot{x}_\theta(0) = x_r(0) = x_\phi(0) = x_\theta(0) = 0$$

Integration of Equation (c) yields

$$\dot{x}_\theta = -2\Omega x_\phi \sin \theta - a\Omega^2 \cos \theta \sin \theta \; t$$

But, except near the maximum altitude, where $\dot{x}_r \simeq 0$, this velocity will be negligible compared with $\dot{x}_r$. Equation b can thus be simplified:

$$\ddot{x}_\phi = -2\Omega \cos \theta \; \dot{x}_r = 2\Omega \cos \theta (v_0 - gt)$$

Integration then yields

$$x_\phi(t) = -\Omega \cos \theta \left(v_0 t^2 - \frac{gt^3}{3} \right)$$

The time in flight may be found by setting $x_r = 0$; this results in

$$t_f = \frac{2v_0}{g}$$

The object will thus return to the earth at a position given by

$$x\left(\frac{2v_0}{g}\right) = -\frac{4}{3}\Omega \cos\theta \frac{v_0{}^3}{g^2}$$

Or, in terms of the maximum height $h = v_0{}^2/2g$,

$$x_\phi\left(2\sqrt{\frac{2h}{g}}\right) = -\frac{8}{3}\Omega \cos\theta\sqrt{\frac{2h^3}{g}}$$

The object thus lands somewhat to the west of the launch position. Observe that this drift results from the Coriolis acceleration component.

Problems

14-34   A horizontal platform rotates around a vertical axis at angular velocity Ω, which gradually increases. A block of mass m is at rest at a distance a from the rotation axis, held there for small Ω by a friction force that is limited by a coefficient of friction μ.
(a) Determine the critical Ω for which the block begins to slide.
(b) Write differential equations of motion that describe the motion of the block after sliding begins, assuming that Ω is approximately constant.

14-35   The pendulum is free to swing in the plane of the uniformly rotating disc. There are no gravitational forces present.
(a) Derive the differential equation that governs the displacement $\phi(t)$.
(b) For $\phi(0) = \phi_0 \ll 1$, $\dot{\phi}(0) = \dot{\phi}_0 \ll \Omega_0$, write the expression for $\phi(t)$.

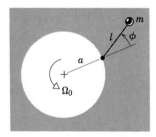

14-36   An object is hurled straight upward, as viewed from an earth-bound reference frame. What will be the $x_r - x_\phi - x_\theta$ components of initial velocity observed from an inertial reference frame translating with the velocity of the center of the earth?

14-37   A shell is fired in a flat trajectory (initial angle of elevation α small) directly northward at high speed, v_0. Neglecting air resistance, what will be the lateral drift due to Coriolis acceleration, in terms of the range b? *Ans.* $\Omega \sin\theta\sqrt{2\alpha b^3/g}$.

14-38 The shell of the previous problem is fired northward from the sixtieth parallel at 6000 ft/s at 8° from the horizontal. What will be the range and the Coriolis drift?

14-39 The block of mass m is constrained to slide along the groove in the horizontal turntable, which is rotating at a constant rate Ω. Springs with a combined stiffness of k (force per deflection) tend to keep the block at the center of the turntable. Friction between the block and the bottom of the groove is negligible; the coefficient of friction between the block and the sides of the groove is μ. Derive the differential equation that governs the displacement $x(t)$.

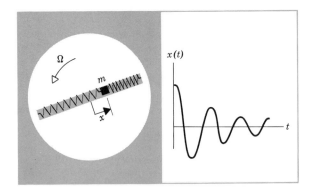

Ans. $\ddot{x} + 2\,\mu|\Omega|\dot{x} + [(k/m) - \Omega^2]x = 0$. Differential equations of this same form arise in the analysis of many physical systems. A procedure for integrating it will be explained in Chapter 18. Without a specific expression for $x(t)$ available, we can expect that for $k/m > \Omega^2$ and relatively small μ the motion is as in the curve.

14-40 Suppose the groove of the previous problem is offset from the center of the disc by the distance h. What will be the equation of motion for x in this case?

14-41 Estimate the inertial-observed acceleration (exclusive of that involved in orbiting around the sun) of an airplane flying due south at an altitude of 5 miles and a constant speed relative to the ground of 1200 mi/h when it crosses its forty-fifth parallel of latitude. Take the radius of the earth to be approximately 4000 miles. Report your findings on a sketch showing *clearly* all components involved.

14-42 Relative to a reference frame α, a force $\mathbf{f}$ is constant. Reference frame β rotates relative to α at angular velocity $\boldsymbol{\Omega}$, which is fixed and perpendicular to $\mathbf{f}$. For the time inerval $t_2 - t_1 = 2\pi/\Omega$, evaluate the α-observed impulse of the force and the β-observed impulse of the force.

14-43   Let us define β-observed work of a force $\mathbf{f}$ as

$$_{\beta}W \;=\; \int \mathbf{f} \cdot d_{\beta}\mathbf{r}$$

where $d_{\beta}\mathbf{r}$ is the β-observed change in position of the particle, and β-observed kinetic energy as

$$_{\beta}T \;=\; \tfrac{1}{2}\,m_{\beta}\mathbf{v} \cdot {}_{\beta}\mathbf{v}$$

Derive the relationship for the β-observed work and kinetic energy, in terms of the velocity of a point B fixed in the moving β reference frame, and its angular velocity $\mathbf{\Omega}$.

14-4

GENERAL MOTION OF A RIGID OBJECT. In Chapter 10 it was pointed out that if a point Q on a rigid body has velocity $\mathbf{v}_Q$, and the body is spinning about an axis through Q, then the velocity of any point P on the body is given by

$$\mathbf{v}_P \;=\; \mathbf{v}_Q + \mathbf{\omega} \times \mathbf{r}_{P/Q}$$

in which the vector $\mathbf{\omega}$ is parallel to the axis of rotation, and has magnitude equal to the rate of rotation and sense determined by the sense of the rotation according to the right-hand rule.

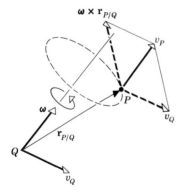

The Angular Velocity Vector.   It is natural to ask whether the motion of a rigid body can always be characterized as a simultaneous translation and rotation in the above manner, or whether inherently more complicated motions are possible. This question is answered by the following *theorem:*

For any motion of a rigid body, there exists a unique vector $\boldsymbol{\omega}$, *called the angular velocity of the body, with the property that the velocity difference between every pair of points on the body is given by*

$$\mathbf{v}_P - \mathbf{v}_Q = \mathbf{v}_{P/Q} = \boldsymbol{\omega} \times \mathbf{r}_{P/Q} \tag{14-7}$$

In order to prove this, we note first that because the body is rigid, the distance between every pair of points is fixed:

$$|\mathbf{r}_{P/Q}| = \sqrt{\mathbf{r}_{P/Q} \cdot \mathbf{r}_{P/Q}} = \text{constant}$$

Differentiation yields the following relationship

$$\mathbf{r}_{P/Q} \cdot \mathbf{v}_{P/Q} = 0$$

Now let the velocities of three noncoplanar points A, B, and C on the body be arbitrary, except for the rigidity constraints

$$\mathbf{r}_{B/C} \cdot \mathbf{v}_{B/C} = 0 \tag{a}$$

$$\mathbf{r}_{C/A} \cdot \mathbf{v}_{C/A} = 0 \tag{b}$$

$$\mathbf{r}_{A/B} \cdot \mathbf{v}_{A/B} = 0 \tag{c}$$

Given $\mathbf{r}_{B/C}$ and $\mathbf{v}_{B/C}$ it is always possible to select a vector $\boldsymbol{\omega}$ that will satisfy

$$\mathbf{v}_{B/C} = \boldsymbol{\omega} \times \mathbf{r}_{B/C} \tag{d}$$

and hence also satisfy the constraint (a). Moreover, we see from Figure 14-9 that there is some flexibility in this choice. Of these infinitely many possibilities

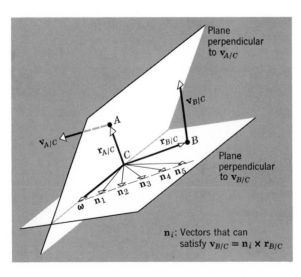

Figure 14-9

Plane perpendicular to $\mathbf{v}_{A/C}$

$\mathbf{v}_{B/C}$

$\mathbf{v}_{A/C}$

A

$\mathbf{r}_{A/C}$ C $\mathbf{r}_{B/C}$ B

Plane perpendicular to $\mathbf{v}_{B/C}$

$\mathbf{n}_3$ $\mathbf{n}_4$ $\mathbf{n}_5$

$\boldsymbol{\omega}$ $\mathbf{n}_1$ $\mathbf{n}_2$

$\mathbf{n}_i$: Vectors that can satisfy $\mathbf{v}_{B/C} = \mathbf{n}_i \times \mathbf{r}_{B/C}$

in the plane perpendicular to $\mathbf{v}_{B/C}$ let us select the one that is also perpendicular to $\mathbf{v}_{A/C}$, so that it also satisfies

$$\boldsymbol{\omega} \cdot \mathbf{v}_{A/C} = 0 \tag{e}$$

Because $\boldsymbol{\omega}$ is perpendicular to $\mathbf{v}_{B/C}$ and $\mathbf{v}_{A/C}$, it must be in the direction of the vector $\mathbf{v}_{B/C} \times \mathbf{v}_{A/C}$. Using (d) and (e), and the vector identity (3-13), we can write this as

$$\mathbf{v}_{B/C} \times \mathbf{v}_{A/C} = (\boldsymbol{\omega} \times \mathbf{r}_{B/C}) \times \mathbf{v}_{A/C}$$

$$= (\boldsymbol{\omega} \cdot \mathbf{v}_{A/C})\mathbf{r}_{B/C} - (\mathbf{v}_{A/C} \cdot \mathbf{r}_{B/C})\boldsymbol{\omega}$$

$$= - (\mathbf{v}_{A/C} \cdot \mathbf{r}_{B/C})\boldsymbol{\omega}$$

The vector $\boldsymbol{\omega}$ is then given as

$$\boldsymbol{\omega} = \frac{\mathbf{v}_{A/C} \times \mathbf{v}_{B/C}}{\mathbf{v}_{A/C} \cdot \mathbf{r}_{B/C}} \tag{f}$$

But the numerator in (f) may be written as

$$\mathbf{v}_{A/C} \times \mathbf{v}_{B/C} = (\mathbf{v}_A - \mathbf{v}_C) \times (\mathbf{v}_B - \mathbf{v}_C)$$

$$= \mathbf{v}_A \times \mathbf{v}_B + \mathbf{v}_B \times \mathbf{v}_C + \mathbf{v}_C \times \mathbf{v}_A$$

and the denominator in (f) may be written as

$$\mathbf{v}_{A/C} \cdot \mathbf{r}_{B/C} = \mathbf{v}_{A/C} \cdot (\mathbf{r}_{B/A} - \mathbf{r}_{C/A})$$

$$= \mathbf{v}_{A/C} \cdot \mathbf{r}_{B/A} + \mathbf{v}_{C/A} \cdot \mathbf{r}_{C/A}$$

$$= \mathbf{v}_{C/A} \cdot \mathbf{r}_{A/B}$$

$$= (\mathbf{v}_{C/B} - \mathbf{v}_{A/B}) \cdot \mathbf{r}_{A/B}$$

$$= \mathbf{v}_{C/B} \cdot \mathbf{r}_{A/B}$$

or, with similar substitutions, as $\mathbf{v}_{B/A} \cdot \mathbf{r}_{C/A}$. Thus, the vector $\boldsymbol{\omega}$ may be expressed as

$$\boxed{\boldsymbol{\omega} = \frac{\mathbf{v}_A \times \mathbf{v}_B + \mathbf{v}_B \times \mathbf{v}_C + \mathbf{v}_C \times \mathbf{v}_A}{vr}} \tag{14-8a}$$

in which the denominator may be written as

$$\boxed{vr = \mathbf{v}_{A/C} \cdot \mathbf{r}_{B/C} = \mathbf{v}_{B/A} \cdot \mathbf{r}_{C/A} = \mathbf{v}_{C/B} \cdot \mathbf{r}_{A/B}} \tag{14-8b}$$

We observe that this expression is completely symmetric in the three subscripts A, B, and C. It follows from this that $\boldsymbol{\omega}$, constructed to satisfy the relationships $\mathbf{v}_{B/C} = \boldsymbol{\omega} \times \mathbf{r}_{B/C}$ and $\boldsymbol{\omega} \perp \mathbf{v}_{A/C}$, also satisfies $\mathbf{v}_{C/A} = \boldsymbol{\omega} \times \mathbf{r}_{C/A}$ and $\mathbf{v}_{A/B} = \boldsymbol{\omega} \times$

$\mathbf{r}_{A/B}$. Thus we have determined a vector $\boldsymbol{\omega}$ that gives each of the velocity differences between pairs among A, B, and C, in the form (14-7).

Now consider a point P other than one of those used in the determination of $\boldsymbol{\omega}$. Its velocity must satisfy the constraints

$$\mathbf{v}_{P/A} \cdot \mathbf{r}_{P/A} = 0$$

$$\mathbf{v}_{P/B} \cdot \mathbf{r}_{P/B} = 0$$

$$\mathbf{v}_{P/C} \cdot \mathbf{r}_{P/C} = 0$$

which, in general, serve to uniquely specify the velocity of point P in terms of the velocities of points A, B, and V. But the relationships

$$\mathbf{v}_{P/i} = \boldsymbol{\omega} \times \mathbf{r}_{P/i} \qquad i = A,B,C$$

satisfy all of these constraints, so must give the values of these velocity differences. Thus, the velocity difference between *any* two points P and Q must be given by

$$\begin{aligned} \mathbf{v}_{P/Q} &= \mathbf{v}_{P/A} - \mathbf{v}_{Q/A} \\ &= \boldsymbol{\omega} \times \mathbf{r}_{P/A} - \boldsymbol{\omega} \times \mathbf{r}_{Q/A} \\ &= \boldsymbol{\omega} \times (\mathbf{r}_{P/A} - \mathbf{r}_{P/A}) \\ &= \boldsymbol{\omega} \times \mathbf{r}_{P/Q} \end{aligned}$$

This proof of the angular velocity theorem has produced the general formula (14-8), which may be used to determine the angular velocity of a body in terms of the velocities of three points. If the velocities of the three points happen to be parallel, the formula becomes indeterminate; in this case the angular velocity may be determined from*

$$\boldsymbol{\omega} = \frac{(\mathbf{v}_{B/C} \cdot \mathbf{v}_{B/C})\mathbf{r}_{A/C}}{(\mathbf{r}_{B/C} \times \mathbf{v}_{B/C}) \cdot \mathbf{r}_{A/C}} + \frac{(\mathbf{v}_{A/C} \cdot \mathbf{v}_{A/C})\mathbf{r}_{B/C}}{(\mathbf{r}_{A/C} \times \mathbf{v}_{A/C}) \cdot \mathbf{r}_{B/C}} \qquad (14\text{-}8c)$$

Example

The universal joint shown in Figure 14-10 is used to transmit rotation from one shaft to another through a bend. One of its features, often undesirable, is that a uniform rotation of one shaft produces a varying speed in the other. Determine

* See Problem 14-52.

Figure 14-10

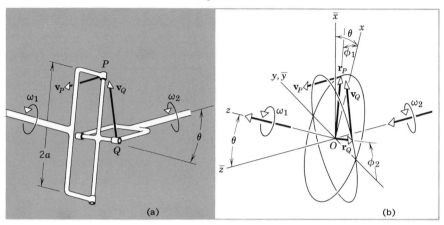

(a)   (b)

the extent of this fluctuation in speed. Also, evaluate the angular velocity of the cross member that connects the two shafts.

Referring to Figure 14-10*b*, we evaluate the velocities of points P and Q in terms of the shaft rotations as follows:

$$\mathbf{r}_P = a(\cos \phi_1 \, \mathbf{u}_x + \sin \phi_1 \, \mathbf{u}_y)$$

$$\mathbf{v}_P = \boldsymbol{\omega}_1 \times \mathbf{r}_P$$

$$= a\omega_1(- \sin \phi_1 \, \mathbf{u}_x + \cos \phi_1 \, \mathbf{u}_y)$$

$$\mathbf{r}_Q = a(\sin \phi_2 \, \bar{\mathbf{u}}_x - \cos \phi_2 \, \bar{\mathbf{u}}_y)$$

$$\mathbf{v}_Q = \boldsymbol{\omega}_2 \times \mathbf{r}_Q$$

$$= a\omega_2(\cos \phi_2 \, \bar{\mathbf{u}}_x + \sin \phi_2 \, \bar{\mathbf{u}}_y)$$

Now, since the cross link is rigid,

$$\mathbf{v}_{P/Q} \cdot \mathbf{r}_{P/Q} = (\mathbf{v}_P - \mathbf{v}_Q) \cdot (\mathbf{r}_P - \mathbf{r}_Q)$$

$$= - (\mathbf{v}_P \cdot \mathbf{r}_Q + \mathbf{v}_Q \cdot \mathbf{r}_P)$$

$$= 0$$

With reference to the figure and the above expressions for the positions and velocities of P and Q, this last equation may be written as

$$a^2\omega_2[\cos \theta \cos \phi_1 \cos \phi_2 + \sin \phi_1 \sin \phi_2]$$

$$- a^2\omega_1[\cos \theta \sin \phi_1 \sin \phi_2 + \cos \phi_1 \cos \phi_2] = 0 \quad \text{(a)}$$

Also, since $\mathbf{r}_P$ and $\mathbf{r}_Q$ are perpendicular,

$$\mathbf{r}_P \cdot \mathbf{r}_Q = a^2(\cos \theta \cos \phi_1 \sin \phi_2 - \sin \phi_1 \cos \phi_2) = 0 \quad \text{(b)}$$

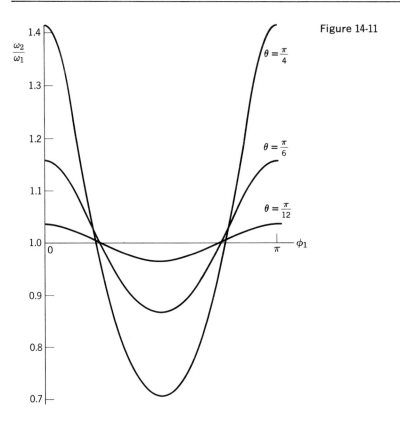

Figure 14-11

This relationship may be used to eliminate ϕ_2 from Equation a, with the result

$$\frac{\omega_2}{\omega_1} = \frac{\cos \theta}{1 - \sin^2 \theta \cos^2 \phi_1} \tag{c}$$

The fluctuations in speed given by this equation are shown in Figure 14-11.

To determine the angular velocity of the cross link, we can apply Equation 14-8, using velocities of points O, P, and Q:

$$\boldsymbol{\omega} = \frac{\mathbf{v}_Q \times \mathbf{v}_P}{\mathbf{v}_Q \cdot \mathbf{r}_P}$$

$$= \frac{a\omega_2(\cos \phi_2\, \bar{\mathbf{u}}_x + \sin \phi_2\, \bar{\mathbf{u}}_y) \times a\omega_1(-\sin \phi_1\, \mathbf{u}_x + \cos \phi_1\, \mathbf{u}_y)}{a^2\omega_2(\cos \theta \cos \phi_1 \cos \phi_2 + \sin \phi_1 \sin \phi_2)}$$

$$= \frac{\sin \phi_1 \sin \phi_2\, \mathbf{u}_z + \cos \phi_1 \sin \phi_2\, \bar{\mathbf{u}}_z - \sin \theta \sin \phi_1 \cos \phi_2\, \mathbf{u}_y}{\cos \theta \cos \phi_1 \cos \phi_2 + \sin \phi_1 \sin \phi_2}$$

Using Equations a and b above, this may be simplified to the form

$$\boldsymbol{\omega} = \tfrac{1}{2}[\omega_1(1 - \cos 2\phi_2)\mathbf{u}_z + \omega_2(1 + \cos 2\phi_1)\bar{\mathbf{u}}_z + \omega_1 \sin \theta \sin 2\phi_2 \, \mathbf{u}_y] \quad \text{(d)}$$

Examination of this equation reveals that as ϕ_1 and ϕ_2 progress from zero to $\pi/2$, the angular velocity vector begins in coincidence with the right-hand shaft, moves out of the plane of the two shafts, and returns to this plane in coincidence with the left-hand shaft. During the subsequent quarter cycle, it moves out on the opposite side of the plane and returns to coincide once again with the left-hand shaft.

Composition of Angular Velocities.   The development of the previous section gives us a means of determining the angular velocity of a body in terms of the velocities of three points on the body. In many situations, rather than knowing the velocities of points on the body, we may know something about the angular velocity of the body relative to another rotating body. To deal with such problems, it is helpful to have a means for determining the angular velocity that corresponds to two or more such relative angular velocities.

To this end, consider the bodies β' and β shown in Figure 14-12. Point B' is fixed in body β', which is rotating with angular velocity $_\alpha\boldsymbol{\omega}_{\beta'}$, with respect to the α reference frame. Points B and P are fixed in body β, which is rotating with angular velocity $_{\beta'}\boldsymbol{\omega}_\beta$ with respect to body β'.

From Equation 14-7, we may relate the β'-observed velocities of points P and B to the relative angular velocity by the equation

$$_{\beta'}\mathbf{v}_P - _{\beta'}\mathbf{v}_B = _{\beta'}\boldsymbol{\omega}_\beta \times \mathbf{r}_{P/B} \quad \text{(a)}$$

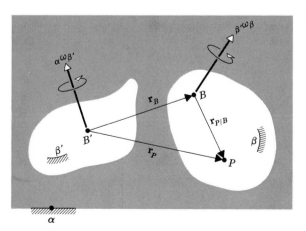

Figure 14-12

Also, the α-observed velocities of points P and B can be related to their β'-observed velocities by means of Equation 14-4:

$$_{\alpha}\mathbf{v}_P = {}_{\beta'}\mathbf{v}_P + {}_{\alpha}\mathbf{v}_{B'} + {}_{\alpha}\boldsymbol{\omega}_{\beta'} \times \mathbf{r}_P \qquad \text{(b)}$$

$$_{\alpha}\mathbf{v}_B = {}_{\beta'}\mathbf{v}_B + {}_{\alpha}\mathbf{v}_{B'} + {}_{\alpha}\boldsymbol{\omega}_{\beta'} \times \mathbf{r}_B \qquad \text{(c)}$$

Subtraction of (c) from (b) results in

$$_{\alpha}\mathbf{v}_P - {}_{\alpha}\mathbf{v}_B = {}_{\beta'}\mathbf{v}_P - {}_{\beta'}\mathbf{v}_B + {}_{\alpha}\boldsymbol{\omega}_{\beta'} \times (\mathbf{r}_P - \mathbf{r}_B)$$

which, on substitution of (a), may be rewritten as

$$_{\alpha}\mathbf{v}_P - {}_{\alpha}\mathbf{v}_B = ({}_{\alpha}\boldsymbol{\omega}_{\beta'} + {}_{\beta'}\boldsymbol{\omega}_{\beta}) \times \mathbf{r}_{P/B} \qquad \text{(d)}$$

But the angular velocity of β relative to α is defined according to Equation 14-7 as the unique vector ${}_{\alpha}\boldsymbol{\omega}_{\beta}$ satisfying

$$_{\alpha}\mathbf{v}_P - {}_{\alpha}\mathbf{v}_B = {}_{\alpha}\boldsymbol{\omega}_{\beta} \times \mathbf{r}_{P/B}$$

for all P and B. Comparing this with Equation d, we see that the angular velocities may be added as follows:

$$\boxed{{}_{\alpha}\boldsymbol{\omega}_{\beta} = {}_{\alpha}\boldsymbol{\omega}_{\beta'} + {}_{\beta'}\boldsymbol{\omega}_{\beta}} \qquad \text{(14-9)}$$

Example

The flywheel in Figure 14-13 is spinning at 100 rad/s in the gimbal frame, which is rotating around the vertical at 25 rad/s. Determine the angular velocity of the flywheel.

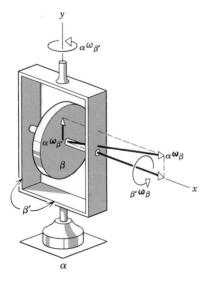

Figure 14-13

With the gimbal frame denoted as β' and the flywheel as β, Equation 14-9 gives the α-observed angular velocity of the flywheel as

$$_\alpha\omega_\beta = 100 \text{ rad/s } \mathbf{u}_x + 25 \text{ rad/s } \mathbf{u}_y$$

This motion, viewed originally as simultaneous rotations about two different axes, may equally well be viewed as a rotation at 103 rad/s around an axis inclined 0.245 rad above the horizontal. Observe how the contributions from $_{\beta'}\omega_\beta$ and $_\alpha\omega_{\beta'}$ combine to produce zero velocity for points along this axis.

This axis is called the *instantaneous axis of rotation* for the flywheel. Note that it remains fixed with respect to the gimbal, and therefore moves with respect to both the ground and the flywheel.

Example

The shaft of the mechanism shown in Figure 14-14 rotates around the vertical at a constant rate of 10 rad/s. The disc is free to rotate around the axis of the shaft, and does not slip at its contact with the horizontal surface. Determine the angular velocity and angular acceleration of the disc.

Considered as an extension of the disc, point O has zero velocity, as has point C'. Hence the line OC' is the instantaneous axis of rotation, defining the direction of the angular velocity of the disc. With the shaft denoted by β' and the disc by β, the vectors appearing in Equation 14-9 are as shown in the figure. The magnitude of the resultant angular velocity is

$$\omega = \left| _\alpha\omega_{\beta'} \right| \text{ ctn } \theta$$

$$= (10 \text{ rpm}) \frac{16 \text{ in.}}{12 \text{ in.}}$$

$$= 1.40 \text{ rad/s}$$

Figure 14-14

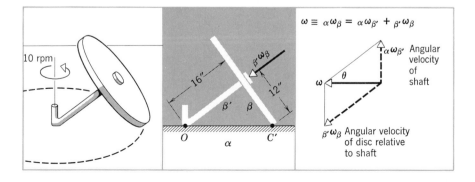

To determine the angular acceleration $\overset{\alpha}{\omega}$, we observe first that the vector ω remains fixed with respect to the shaft. With a reference frame attached to the shaft, an application of Equation 14-1 yields

$$\overset{\alpha}{\omega} = \overset{\beta'}{\omega} + {}_\alpha\omega_{\beta'} \times \omega$$

$$= \mathbf{0} + |{}_\alpha\omega_{\beta'}|\omega \sin \frac{\pi}{2} \, \mathbf{u}_z$$

$$= 1.46 \text{ rad/s}^2 \, \mathbf{u}_z$$

where $\mathbf{u}_z$ is directed toward the viewer of the vector diagram in the figure.

Example

Determine the speed reduction in the bevel-gear train shown in Figure 14-15. In this mechanism, the two gear unit β turns with respect to the arm around the inclined axis, while the arm rotates freely around a horizontal axis.

As a first step, let us determine the motion of the gear unit β. Referring to Figure 14-15b, note that points O and C' on this body have zero velocity; thus the line OC' is its instantaneous axis of rotation and defines the direction of the angular velocity ω_β. But, according to Equation 14-9, this angular velocity is the sum of that of the arm and that of the gear relative to the arm, and the directions of these components are as indicated in Figure 14-15c. Now, the magnitude of $\omega_{\beta'}$, is related to the speed of point Q by

$$v_Q = \left(a + \frac{56}{20} a \sin \phi \right)\omega_{\beta'}$$

$$= \frac{12}{5} a\omega_{\beta'}$$

and since Q lies one-half the distance from the instantaneous axis as does P,

$$v_Q = \tfrac{1}{2} v_P$$

But since gears 1 and β are engaged at P,

$$v_P = a\omega_1$$

These last three relationships give us

$$\omega_{\beta'} = \frac{5}{24} \omega_1$$

Both components $\omega_{\beta'}$ and ${}_{\beta'}\omega_\beta$ contribute to the speed of point C', which, as we have already noted, is zero:

$$v_{C'} = \frac{76}{20} a\omega_{\beta'} - \frac{56}{20} a \, {}_{\beta'}\omega_\beta = 0$$

Figure 14-15

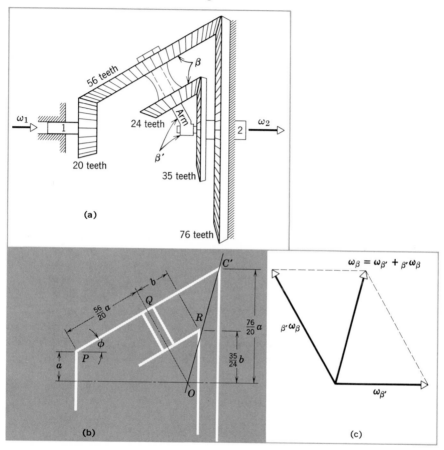

(a)

(b)

(c)

and, therefore,

$$\beta'\omega_\beta = \frac{76}{56}\omega_\beta$$

Now, the speed of point R is determined in the same way:

$$v_R = \frac{35}{24}b\omega_{\beta'} - b_{\,\beta'\omega_\beta}$$

But, since the output gear 2 and β are engaged at R,

$$\omega_2 = \frac{v_R}{\frac{35}{24}b}$$

The above has taken us step by step through the train of moving parts—angular velocity of one part giving the tangential velocity of a common point on the succeeding part, giving the angular velocity of that part, and so on. Algebraic elimination of the intermediate speeds from the above equations gives the output speed directly in terms of the input speed:

$$\omega_2 = \omega_{\beta'} - \frac{24}{35} \beta'\omega_\beta$$

$$= \frac{5}{24} \omega_1 - \frac{24}{35} \frac{76}{56} \frac{5}{24} \omega_1$$

$$= \frac{17}{1176} \omega_1$$

This mechanism, called a Humpage's reduction gear train, achieves a speed reduction of 69.2:1!

Screw Motion.   The preceding examples, along with those in Chapter 10 in which we employed the instantaneous center, suggest that the instantaneous axis might form the basis for an effective approach to every analysis of rigid body kinematics. But this would assume that such an axis always exists.

The question raised here may be answered by examining the resolution of the velocity of a point P into a component parallel to the angular velocity and a component perpendicular to the angular velocity (see Equation 3-15):

$$\mathbf{v}_P = \frac{(\boldsymbol{\omega}\cdot\mathbf{v}_P)\boldsymbol{\omega}}{\boldsymbol{\omega}\cdot\boldsymbol{\omega}} + \frac{\boldsymbol{\omega}\times(\mathbf{v}_P\times\boldsymbol{\omega})}{\boldsymbol{\omega}\cdot\boldsymbol{\omega}}$$

The significance of this equation is more readily understood after it is put in the form

$$\mathbf{v}_P = \mathbf{v}_\omega + \boldsymbol{\omega}\times\boldsymbol{\rho}_{P/C'} \tag{14-10a}$$

in which

$$\mathbf{v}_\omega = \frac{(\boldsymbol{\omega}\cdot\mathbf{v}_P)\boldsymbol{\omega}}{\boldsymbol{\omega}\cdot\boldsymbol{\omega}} \tag{14-10b}$$

$$\boldsymbol{\rho}_{P/C'} = \frac{\mathbf{v}_P\times\boldsymbol{\omega}}{\boldsymbol{\omega}\cdot\boldsymbol{\omega}} \tag{14-10c}$$

Comparison of Equation 14-10a with Equation 14-7 reveals that $\boldsymbol{\rho}_{P/C'}$ locates the point P relative to a point C' that has the resultant velocity $\mathbf{v}_\omega$. Moreover, as can be ascertained from Figure 14-16, every point on the straight line passing through C' and parallel to $\boldsymbol{\omega}$ has the same resultant velocity. This line, called the *screw axis*, is particularly significant because, as Equation 14-10a indicates, every other point on the body is moving at a higher speed.

Figure 14-16

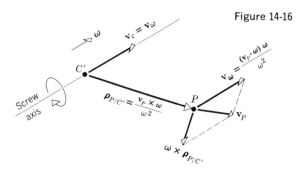

From this we see that an instantaneous axis of rotation (a set of points having zero velocity) will exist if and only if the angular velocity vector is perpendicular to the resultant velocity of every point. And it follows from this that an instantaneous center exists for every planar motion in which $\omega \neq 0$.

It is interesting to compare this composition of rigid body motion into the *screw motion*, with the composition of a system of forces into the *wrench* (Section 3-5).

Example

A helicopter rotor is rotating at 70 rad/s, while the helicopter travels horizontally at 35 m/s. The rotor is tipped forward at the angle χ from the vertical. Considering the rotor as a rigid body, is there an instantaneous axis, and if so, where does it lie?

As shown in Figure 14-17, the resolution of motion given consists of the velocity of the rotor center $\mathbf{v}_P$, and the angular velocity $\boldsymbol{\omega}$. Resolving the

Figure 14-17

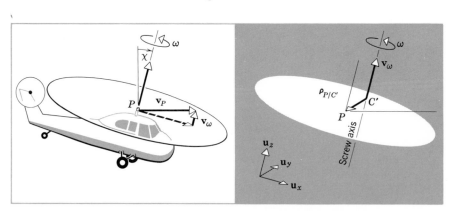

velocity of point P into a component parallel to $\boldsymbol{\omega}$ and a component perpendicular to $\boldsymbol{\omega}$, we find the velocity and location of points on the screw axis:

$$\mathbf{v}_\omega = 35 \sin \chi \text{ m/s } \mathbf{u}_z$$

$$\boldsymbol{\rho}_{P/C'} = \frac{\mathbf{v}_P \times \boldsymbol{\omega}}{\omega^2} = -\frac{v_P \cos \chi}{\omega} \mathbf{u}_y = -0.50 \cos \chi \text{ m } \mathbf{u}_y$$

The motion may thus be viewed as a translation in the direction of the angular velocity vector at the speed $35 \sin \chi$ m/s and a rotation about the screw axis that is located a distance $0.5 \cos \chi$ m from the point P. As long as the rotor is tilted, there are no points with zero velocity; if $\chi = 0$, then $\mathbf{v}_\omega = \mathbf{0}$ and the screw axis becomes an instantaneous axis of rotation.

Problems

14-44   A gun barrel is swinging around the vertical and upward, simultaneously. Internal pressure imparts a force $f(t)$ to the cannonball.
(a) Give the $\mathbf{u}_r - \mathbf{u}_\phi - \mathbf{u}_\theta$ components of the angular velocity of the barrel in terms of ϕ, θ, and their derivatives.
(b) Neglecting friction, write the differential equation that governs $r(t)$.

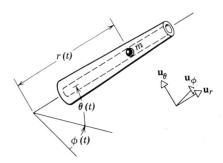

14-45   The cannon of the previous problem is swinging around the vertical at a constant 2 rad/s. The elevation is fixed at 38°. Gravity and friction are negligible. The 3-kg cannonball is at $r = 2$m and traveling at 700 m/s relative to the barrel. Draw a free-body diagram of the cannonball and indicate the forces exerted on it.

14-46   Without evaluating the velocities of points within the mechanism, determine the speed ratio ω_2/ω_1, and the angular velocity of the cross member in the universal joint shown in Figure 14-10. *Suggestion.* Note that the mechanism constrains the directions of the relative angular velocities between each shaft and the cross member.

14-47   Determine the ground-observed angular velocity of the disc, and the velocity and acceleration of the point A on the disc.

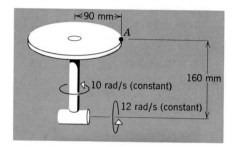

14-48   In order to analyze the motion of the symmetrical satellite, the angles shown have been introduced. Evaluate the angular velocity as seen from an inertial reference frame, giving its components on the x axes.

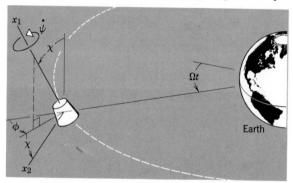

14-49   The fork A rotates with constant angular velocity ω_A. The disc B rotates with constant angular velocity ω_B relative to the fork. The

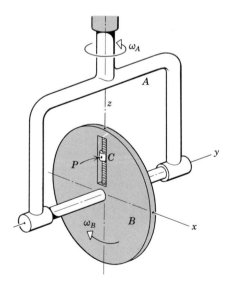

collar C has a diameter d and spins on the threaded shaft that has a pitch diameter e and a pitch angle θ.

(a) Determine the ground-observed velocity and acceleration of a point on the surface of the collar, in the position shown.

(b) Evaluate the ground-observed velocity and acceleration of the point when the axis of the threaded shaft has rotated 90° from the position shown.

14-50 Determine the angular velocity of disc 4.

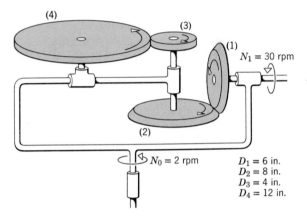

(4) (3) (1) $N_1 = 30$ rpm

(2) $N_0 = 2$ rpm

$D_1 = 6$ in.
$D_2 = 8$ in.
$D_3 = 4$ in.
$D_4 = 12$ in.

14-51 A person is sitting in the seat as it spins about the B-B axis as shown and the arm, in which the B-B axis is fixed, spins about A-A. Estimate the acceleration of the person's head at the instant shown, assuming the angular velocities to be constant.

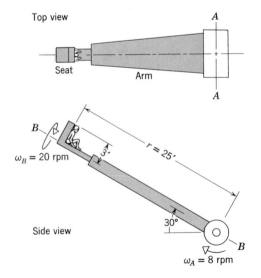

Top view

Seat

Arm

A

A

B

$\omega_B = 20$ rpm

3'

$r = 25'$

30°

Side view

B

$\omega_A = 8$ rpm

14-52   When $\mathbf{v}_{A/C}$ and $\mathbf{v}_{B/C}$ happen to be parallel, the formula (14-8a) becomes indeterminate.

(a) Copy and complete the accompanying sketch to show the location of the instantaneous axis of rotation for $\mathbf{v}_{A/C}$ and $\mathbf{v}_{B/C}$.

(b) Show from this that the angular velocity lies in the plane of $\mathbf{r}_{A/C}$ and $\mathbf{r}_{B/C}$, and hence may be represented in the form

$$\boldsymbol{\omega} = a\mathbf{r}_{A/C} + b\mathbf{r}_{B/C}$$

(c) Verify Equation 14-8c.

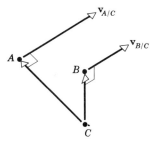

14-53   Show formally that if $\mathbf{v}_{P/Q} = \boldsymbol{\omega} \times \mathbf{r}_{P/Q}$, then $\mathbf{v}_{P/Q} \cdot \mathbf{r}_{P/Q} = 0$.

14-54   A rigid body is moving in a plane in such a way that a point O on the body is stationary. With point P located relative to O by the position vector $\mathbf{r}_P$ and its velocity given by $\mathbf{v}_P$, show that the angular velocity of the body is

$$\boldsymbol{\omega} = \frac{\mathbf{r}_P \times \mathbf{v}_P}{r_P^2}$$

14-55   What would be the reduction ratio in the Humpage's reduction mechanism of Figure 14-15, if the numbers of gear teeth were selected so point R fell on the line OC'?

14-56   The symmetry axis of the moving gear swings around the vertical at a rate ω_P.

(a) Determine the angle ϕ.

(b) Show the direction of the resultant angular velocity of the gear.

(c) Determine the magnitude of the resultant angular velocity in terms of ω_P.

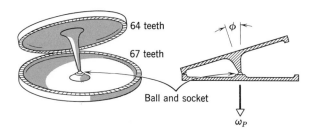

64 teeth

67 teeth

Ball and socket

ω_P

14-57 The shaft swings around the vertical at the angular speed ω_P. A bearing constrains the disc to remain perpendicular to the shaft, but allows it to turn relative to the shaft so that it may roll around the horizontal plane without slip. Evaluate the angular velocity and angular acceleration of the disc.

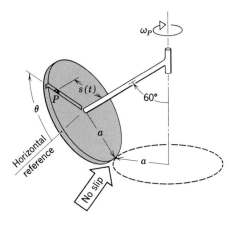

14-58 The disc of the previous problem has a radial groove along which a particle P is constrained to move. The position of P in the groove is prescribed by the distant $s(t)$. Determine the ground-observed velocity and acceleration of P.

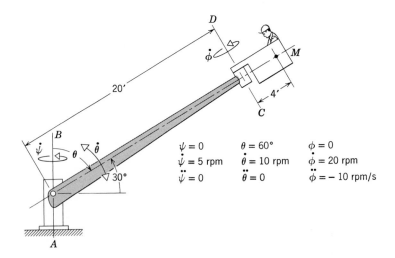

14-59 Estimate the acceleration of the man in the amusement park device depicted in the sketch. The entire assembly rotates about axis A-B as the arm swings up and down and the car whips back and forth about axis C-D. At the instant shown the car extends straight out from the arm and the arm is at a 30° angle with the horizontal. ψ is specified relative to the ground, θ relative to the axis A-B, and ϕ relative to the arm.

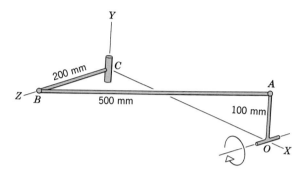

14-60 Arm AO rotates with angular velocity ω_A about an axis parallel to the Z axis. Arm CB rotates about the Y axis. Determine the angular velocity of arm CB.

COORDINATE AND VECTOR TRANSFORMATIONS

The analysis of rigid body motions often requires that we keep track of the interdependence of several sets of vector components. This chapter deals with procedures that permit orderly handling of such interrelationships, which can otherwise present a formidable tangle.

15-1

SOME MATRIX ALGEBRA. One of the most effective tools devised for this task is the algebra of *matrices*. The procedure consists in forming ordered arrays of quantities, and operating with these as single entities, according to rules for equality, addition, multiplication, and so on. This places less interesting detail in the background, bringing out more clearly an overall view of the operations.

Definitions. A *matrix* is defined as an ordered, two-dimensional array of quantities, obeying laws of algebra to be defined in the following paragraphs. An example would be the matrix

$$\begin{bmatrix} 0.940 & 0.342 & 0.000 \\ -0.342 & 0.940 & 0.000 \\ 0.000 & 0.000 & 1.000 \end{bmatrix}$$

This particular matrix has three rows and three columns and is called a *three-dimensional square* matrix. Observe that we enclose the matrix with brackets, which must be clearly distinguished from the vertical bars used to denote a determinant.

A matrix may have any number of rows and columns. One that has only one column is called a column matrix, and one that has only one row is called a row matrix. In order to more readily distinguish them from other matrices, we will enclose column matrices in braces, as, for example*

$$\begin{Bmatrix} 0.342 \\ 0.940 \\ 0.000 \end{Bmatrix}$$

Equality of two matrices implies equality of the members of each pair of elements in corresponding positions in the two matrices.

Addition or *subtraction* results in a matrix having each element equal to the sum or difference formed from the corresponding elements of the matrices entering the operation.

The *product* of a *scalar* and a *matrix* is a matrix having each element equal to the product of the scalar and the corresponding element in the matrix being multiplied.

The rule for *multiplication* of two matrices is as follows:

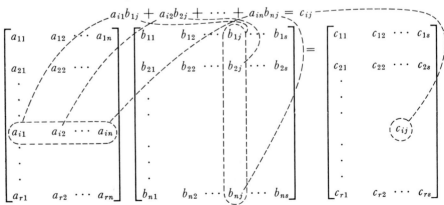

* This convention is fairly common, but by no means universal. Many writers use square brackets [ ], curved brackets ( ), or double lines ‖   ‖ to enclose all matrices.

The following examples illustrate this process:

$$(5)(0) + (1)(2) + (-3)(-3) = 11$$

$$\begin{bmatrix} 1 & -1 & 2 \\ 1 & 0 & 3 \\ 5 & 1 & -3 \end{bmatrix} \begin{bmatrix} 0 & 1 & 2 \\ 2 & 2 & 0 \\ -3 & -1 & -2 \end{bmatrix} = \begin{bmatrix} -8 & -3 & -2 \\ -9 & -2 & -4 \\ 11 & 10 & 16 \end{bmatrix}$$

$$\begin{bmatrix} 0 & 1 & 1 \\ 2 & 2 & 0 \\ -3 & -1 & -2 \end{bmatrix} \begin{bmatrix} 1 & -1 & 2 \\ 1 & 0 & 3 \\ 5 & 1 & -3 \end{bmatrix} = \begin{bmatrix} 6 & 1 & 0 \\ 4 & -2 & 10 \\ -14 & 1 & -3 \end{bmatrix}$$

$$\begin{bmatrix} 0 & 1 & 1 \\ 2 & 2 & 0 \end{bmatrix} \begin{bmatrix} 1 & -1 \\ 1 & 0 \\ 5 & 1 \end{bmatrix} = \begin{bmatrix} 6 & 1 \\ 4 & -2 \end{bmatrix}$$

$$\begin{bmatrix} 1 & -1 \\ 1 & 0 \\ 5 & 1 \end{bmatrix} \begin{bmatrix} 0 & 1 & 1 \\ 2 & 2 & 0 \\ -3 & -1 & -2 \end{bmatrix}$$ is nonsense. These are said to be not *conformable* in this order.

$$\begin{bmatrix} 0 & 1 & 1 \\ 2 & 2 & 0 \\ -3 & -1 & -2 \end{bmatrix} \begin{Bmatrix} 1 \\ 1 \\ 5 \end{Bmatrix} = \begin{Bmatrix} 6 \\ 4 \\ -14 \end{Bmatrix}$$

$$\begin{bmatrix} 0 & 1 & 1 \\ 2 & 2 & 0 \end{bmatrix} \begin{Bmatrix} 1 \\ 1 \\ 5 \end{Bmatrix} = \begin{Bmatrix} 6 \\ 4 \end{Bmatrix}$$

$$\begin{bmatrix} 0 & 1 & 1 \end{bmatrix} \begin{Bmatrix} 1 \\ 1 \\ 5 \end{Bmatrix} = 6 \qquad \begin{Bmatrix} 1 \\ 1 \\ 5 \end{Bmatrix} \begin{bmatrix} 0 & 1 & 1 \end{bmatrix} = \begin{bmatrix} 0 & 1 & 1 \\ 0 & 1 & 1 \\ 0 & 5 & 5 \end{bmatrix}$$

In addition to giving practice in the mechanics of multiplication, study of the above examples will point up the following facts: The matrix resulting from the multiplication has the same number of rows as has the left-hand factor (the *prefactor*) and the same number of columns as has the right-hand factor (the *postfactor*). If multiplication is to be carried out at all, the number of columns in the prefactor must be equal to the number of rows in the postfactor. Unlike the multiplication of ordinary numbers, the commutative law does not hold for multiplication of matrices; that is, the order of multiplication affects the resulting product. However, it can be shown that the associative and distributive laws do hold.

The *transpose* of a matrix is defined as the matrix resulting from interchange of the rows and columns of the original matrix. We denote this with a superscript T; thus,

$$\begin{bmatrix} 2 & 0 & 3 \\ 4 & 1 & -6 \\ -2 & 2 & 5 \end{bmatrix}^T = \begin{bmatrix} 2 & 4 & -2 \\ 0 & 1 & 2 \\ 3 & -6 & 5 \end{bmatrix}$$

$$\begin{Bmatrix} A_x \\ A_y \\ A_z \end{Bmatrix}^T = [A_x \ A_y \ A_z]$$

A set of *simultaneous linear algebraic* equations, such as

$$3x_1 - 2x_2 = y_1$$
$$-x_1 + x_2 = y_2$$

may be represented by an equivalent matrix equation, such as

$$\begin{bmatrix} 3 & -2 \\ -1 & 1 \end{bmatrix} \begin{Bmatrix} x_1 \\ x_2 \end{Bmatrix} = \begin{Bmatrix} y_1 \\ y_2 \end{Bmatrix} \tag{a}$$

By means of Cramer's rule, or another method, these equations may be solved to give x_1 and x_2 in terms of y_1 and y_2:

$$x_1 = y_1 + 2y_2$$
$$x_2 = y_1 + 3y_2$$

Or,

$$\begin{Bmatrix} x_1 \\ x_2 \end{Bmatrix} = \begin{bmatrix} 1 & 2 \\ 1 & 3 \end{bmatrix} \begin{Bmatrix} y_1 \\ y_2 \end{Bmatrix} \tag{b}$$

Equation b can be formally obtained by premultiplying (a) by the matrix
$\begin{bmatrix} 1 & 2 \\ 1 & 3 \end{bmatrix}$:

$$\begin{bmatrix} 1 & 2 \\ 1 & 3 \end{bmatrix}\begin{bmatrix} 3 & -2 \\ -1 & 1 \end{bmatrix}\begin{Bmatrix} x_1 \\ x_2 \end{Bmatrix} = \begin{bmatrix} 1 & 2 \\ 1 & 3 \end{bmatrix}\begin{Bmatrix} y_1 \\ y_2 \end{Bmatrix}$$

$$\begin{bmatrix} 1 & 0 \\ 0 & 1 \end{bmatrix}\begin{Bmatrix} x_1 \\ x_2 \end{Bmatrix} = \begin{bmatrix} 1 & 2 \\ 1 & 3 \end{bmatrix}\begin{Bmatrix} y_1 \\ y_2 \end{Bmatrix}$$

$$\begin{Bmatrix} x_1 \\ x_2 \end{Bmatrix} = \begin{bmatrix} 1 & 2 \\ 1 & 3 \end{bmatrix}\begin{Bmatrix} y_1 \\ y_2 \end{Bmatrix}$$

The square matrix appearing on the left-hand side in the second step is called the *identity matrix*, or *unit matrix*. It is defined in n dimensions as the square matrix*

$$[1] = \begin{bmatrix} 1 & 0 & 0 & \cdots & 0 \\ 0 & 1 & 0 & \cdots & 0 \\ 0 & 0 & 1 & \cdots & 0 \\ & & \vdots & & \\ 0 & 0 & 0 & \cdots & 1 \end{bmatrix}$$

Multiplication of any conformable matrix by the unit matrix leaves the matrix unchanged. Thus [1] plays the same role in matrix algebra that the number 1 plays in ordinary algebra.

The matrices $\begin{bmatrix} 3 & -2 \\ -1 & 1 \end{bmatrix}$ and $\begin{bmatrix} 1 & 2 \\ 1 & 3 \end{bmatrix}$ are the *inverses* of one another, written as

$$\begin{bmatrix} 3 & -2 \\ -1 & 1 \end{bmatrix}^{-1} = \begin{bmatrix} 1 & 2 \\ 1 & 3 \end{bmatrix}$$

$$\begin{bmatrix} 1 & 2 \\ 1 & 3 \end{bmatrix}^{-1} = \begin{bmatrix} 3 & -2 \\ -1 & 1 \end{bmatrix}$$

* Another abbreviation commonly used for the identity matrix is $[I]$ or I.

The *inverse* of a square matrix $[A]$ is defined as the square matrix $[A]^{-1}$ having the property

$$[A]\,[A]^{-1} = [A]^{-1}\,[A] = [1]$$

It can be shown that this uniquely defines the inverse of the matrix $[A]$, provided its determinant is not zero. If $\det[A] = 0$, no inverse exists. The actual determination of the inverse of a given matrix is a fairly tedious process in general. We will not pursue this here, because our applications require computing inverses only for a special class of matrices for which the inverse is equal to the transpose.

Products of Vectors.   A vector may be represented in matrix notation by forming a row or column matrix from its components. Thus, we understand

$$\mathbf{A}: \quad \begin{Bmatrix} A_x \\ A_y \\ A_z \end{Bmatrix}$$

to be the equivalent of

$$\mathbf{A} = A_x\mathbf{u}_x + A_y\mathbf{u}_y + A_z\mathbf{u}_z$$

Or, the equation

$$\begin{Bmatrix} F_x \\ F_y \\ F_z \end{Bmatrix} = \begin{Bmatrix} 30 \text{ kN} \\ 28 \text{ kN} \\ -17 \text{ kN} \end{Bmatrix}$$

to be the equivalent of

$$F_x\mathbf{u}_x + F_y\mathbf{u}_y + F_z\mathbf{u}_z = (30 \text{ kN})\mathbf{u}_x + (28 \text{ kN})\mathbf{u}_y + (-17 \text{ kN})\mathbf{u}_z$$

In terms of rectangular Cartesian components, the dot product of the vectors $\mathbf{A}$ and $\mathbf{B}$ may be written as

$$\mathbf{A}\cdot\mathbf{B} = A_xB_x + A_yB_y + A_zB_z$$

$$= [A_x\,A_y\,A_z]\begin{Bmatrix} B_x \\ B_y \\ B_z \end{Bmatrix}$$

$$= \{A\}^T\{B\} \tag{15-1}$$

where we make no distinction between a scalar and the one-by-one matrix resulting from the multiplication.

We showed earlier that the cross product may be written in terms of rectangular Cartesian components as

$$\mathbf{A} \times \mathbf{B} = (A_y B_z - A_z B_y)\mathbf{u}_x$$
$$+ (A_z B_x - A_x B_z)\mathbf{u}_y$$
$$+ (A_x B_y - A_y B_x)\mathbf{u}_z$$

Or, forming a column matrix from the components of the cross product, we can write the equivalent as

$$\begin{Bmatrix} (\mathbf{A} \times \mathbf{B}) \cdot \mathbf{u}_x \\ (\mathbf{A} \times \mathbf{B}) \cdot \mathbf{u}_y \\ (\mathbf{A} \times \mathbf{B}) \cdot \mathbf{u}_z \end{Bmatrix} = \begin{bmatrix} 0 & -A_z & A_y \\ A_z & 0 & -A_x \\ -A_y & A_x & 0 \end{bmatrix} \begin{Bmatrix} B_x \\ B_y \\ B_z \end{Bmatrix}$$
$$= [A\times]\{B\} \tag{15-2}$$

where the abbreviation $[A\times]$ will be used in this text to mean

$$[A\times] = \begin{bmatrix} 0 & -A_z & A_y \\ A_z & 0 & -A_x \\ -A_y & A_x & 0 \end{bmatrix} \tag{15-2a}$$

Problems

15-1   Verify the results on p. 51.

15-2   Verify that Equations 15-1 and 15-2 correctly represent the dot and cross products.

15-3   Referring to Problem 14-13, show that the unit vectors in the tangent, normal, and binormal directions vary according to

$$\frac{d}{ds} \begin{Bmatrix} \mathbf{u}_t \\ \mathbf{u}_n \\ \mathbf{u}_b \end{Bmatrix} = \begin{bmatrix} 0 & \kappa & 0 \\ -\kappa & 0 & \tau \\ 0 & -\tau & 0 \end{bmatrix} \begin{Bmatrix} \mathbf{u}_t \\ \mathbf{u}_n \\ \mathbf{u}_b \end{Bmatrix}$$

15-4   Verify that the distributive law holds for the examples

$$\begin{bmatrix} 2 & 1 \\ -3 & 6 \end{bmatrix} \begin{bmatrix} \begin{bmatrix} 1 & -2 \\ -2 & 3 \end{bmatrix} + \begin{bmatrix} 5 & 3 \\ 2 & -1 \end{bmatrix} \end{bmatrix}$$

$$\left[\begin{array}{cc} 1 & -2 \\ -2 & 3 \end{array}\right] + \left[\begin{array}{cc} 5 & 3 \\ 2 & -1 \end{array}\right]\left[\begin{array}{cc} 2 & 1 \\ -3 & 6 \end{array}\right]$$

15-5 Demonstrate the validity of the distributive law of matrix multiplication in general.

15-6 Show how to write the three matrix equations in a single matrix equation.

$$\{D\} = [T]\{A\}$$
$$\{E\} = [T]\{B\}$$
$$\{F\} = [T]\{C\}$$

where $[T]$ is a square matrix.

15-7 Express the differential equations

$$3\ddot{x} + 6x - 2y = f_x(t)$$
$$2\ddot{y} - 2x + 4y = f_y(t)$$

in matrix notation:

$$\left[\begin{array}{cc} & \\ & \end{array}\right]\left\{\begin{array}{c} \ddot{x} \\ \ddot{y} \end{array}\right\} + \left[\begin{array}{cc} & \\ & \end{array}\right]\left\{\begin{array}{c} x \\ y \end{array}\right\} = \left\{\begin{array}{c} \\ \end{array}\right\}$$

15.8 Evaluate the inverse of the matrix

$$\left[\begin{array}{cccc} 1 & 0 & 0 & 0 \\ 0 & 2 & 0 & 0 \\ 0 & 0 & -3 & 0 \\ 0 & 0 & 0 & 5 \end{array}\right]$$

15-9 Solve the simultaneous equations for x_1 and x_2:

$$2x_1 + 3x_2 = y_1$$
$$4x_1 + x_2 = y_2$$

Write the inverse,

$$\left[\begin{array}{cc} 2 & 3 \\ 4 & 1 \end{array}\right]^{-1} = \left[\begin{array}{cc} & \\ & \end{array}\right]$$

15-10 Evaluate the product

$$[B_x \; B_y \; B_z] \begin{bmatrix} 0 & -A_z & A_y \\ A_z & 0 & -A_x \\ -A_y & A_x & 0 \end{bmatrix} \begin{Bmatrix} B_x \\ B_y \\ B_z \end{Bmatrix}$$

In terms of what you know about vectors, how could you have pre-dicted the result?

15-11 Show that if the matrix $[A]$ is *symmetric* (i.e., $A_{ij} = A_{ji}$) that

$$\{x\}^T[A]\{y\} = \{y\}^T[A]\{x\}$$

15-2

RIGID ROTATION OF AXES. Consider the rotation of the rectangular Cartesian coordinate axes shown in Figure 15-1. The orientation of the $\bar{x}, \bar{y}, \bar{z}$ axes relative to the x, y, z axes is specified by the nine *direction cosines*

$$l_{\bar{x}x} = \cos \measuredangle{}^x_{\bar{x}} \qquad l_{\bar{x}y} = \cos \measuredangle{}^y_{\bar{x}} \qquad l_{\bar{x}z} = \cos \measuredangle{}^z_{\bar{x}}$$

$$l_{\bar{y}x} = \cos \measuredangle{}^x_{\bar{y}} \qquad l_{\bar{y}y} = \cos \measuredangle{}^y_{\bar{y}} \qquad l_{\bar{y}z} = \cos \measuredangle{}^z_{\bar{y}}$$

$$l_{\bar{z}x} = \cos \measuredangle{}^x_{\bar{z}} \qquad l_{\bar{z}y} = \cos \measuredangle{}^y_{\bar{z}} \qquad l_{\bar{z}z} = \cos \measuredangle{}^z_{\bar{z}}$$

A vector **A** may be expressed by giving its components in either coordinate system, as

$$\mathbf{A} = A_x\mathbf{u}_x + A_y\mathbf{u}_y + A_z\mathbf{u}_z \qquad\qquad (15\text{-}3a)$$

$$\mathbf{A} = \bar{A}_x\bar{\mathbf{u}}_x + \bar{A}_y\bar{\mathbf{u}}_y + \bar{A}_z\bar{\mathbf{u}}_z \qquad\qquad (15\text{-}3b)$$

Dot multiplication of the first of these equations with $\bar{\mathbf{u}}_x$, $\bar{\mathbf{u}}_y$, then $\bar{\mathbf{u}}_z$ results in

$$\mathbf{A}\cdot\bar{\mathbf{u}}_x = A_x\bar{\mathbf{u}}_x\cdot\mathbf{u}_x + A_y\bar{\mathbf{u}}_x\cdot\mathbf{u}_y + A_z\bar{\mathbf{u}}_x\cdot\mathbf{u}_z$$

$$\mathbf{A}\cdot\bar{\mathbf{u}}_y = A_x\bar{\mathbf{u}}_y\cdot\mathbf{u}_x + A_y\bar{\mathbf{u}}_y\cdot\mathbf{u}_y + A_z\bar{\mathbf{u}}_y\cdot\mathbf{u}_z$$

$$\mathbf{A}\cdot\bar{\mathbf{u}}_z = A_x\bar{\mathbf{u}}_z\cdot\mathbf{u}_x + A_y\bar{\mathbf{u}}_z\cdot\mathbf{u}_y + A_z\bar{\mathbf{u}}_z\cdot\mathbf{u}_z$$

But the geometric meanings of quantities in these equations are expressed by

$$\bar{\mathbf{u}}_i\cdot\mathbf{A} = \bar{A}_i \qquad\qquad (i = x,y,z)$$

$$\bar{\mathbf{u}}_i\cdot\mathbf{u}_j = |\bar{\mathbf{u}}_i|\,|\mathbf{u}_j|\cos \measuredangle{}^j_{\bar{i}} = l_{\bar{i}j} \qquad (i,j = x,y,z)$$

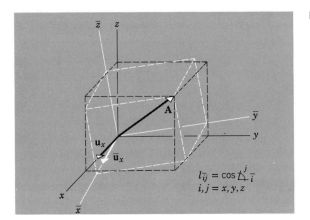

Figure 15-1

$$l_{\bar{i}j} = \cos \angle_{\,\bar{i}}^{\,j}$$
$$i, j = x, y, z$$

Therefore, the above equations give us the relationships among the components of the vector in the two coordinate systems, in terms of the direction cosines that define the transformation:

$$\bar{A}_x = l_{\bar{x}x}A_x + l_{\bar{x}y}A_y + l_{\bar{x}z}A_z$$

$$\bar{A}_y = l_{\bar{y}x}A_x + l_{\bar{y}y}A_y + l_{\bar{y}z}A_z$$

$$\bar{A}_z = l_{\bar{z}x}A_x + l_{\bar{z}y}A_y + l_{\bar{z}z}A_z$$

Or, in matrix notation,

$$\begin{Bmatrix} \bar{A}_x \\ \bar{A}_y \\ \bar{A}_z \end{Bmatrix} = \begin{bmatrix} l_{\bar{x}x} & l_{\bar{x}y} & l_{\bar{x}z} \\ l_{\bar{y}x} & l_{\bar{y}y} & l_{\bar{y}z} \\ l_{\bar{z}x} & l_{\bar{z}y} & l_{\bar{z}z} \end{bmatrix} \begin{Bmatrix} A_x \\ A_y \\ A_z \end{Bmatrix} \tag{15-4}$$

Example

A vector has components (2,1,3) in a rectangular Cartesian coordinate system. What will be its components relative to a coordinate system that is obtained from the first by a 30° right-handed rotation about the z axis?

Examination of the figure reveals the following values for the direction cosines:

$$l_{\bar{x}x} = \cos 30° \qquad l_{\bar{x}y} = \cos 60° \qquad l_{xz} = \cos 90°$$
$$= 0.866 \qquad\qquad = 0.500 \qquad\qquad = 0$$

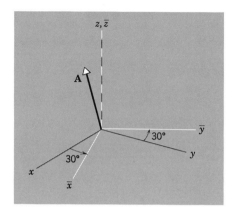

$$l_{\bar{y}x} = \cos 120° \qquad l_{\bar{y}y} = \cos 30° \qquad l_{\bar{y}z} = \cos 90°$$
$$\qquad = -0.500 \qquad\qquad = 0.866 \qquad\qquad = 0$$

$$l_{\bar{z}x} = \cos 90° \qquad l_{\bar{z}y} = \cos 90° \qquad l_{\bar{z}z} = \cos 0°$$
$$\qquad = 0 \qquad\qquad\quad = 0 \qquad\qquad\quad = 1$$

Next, the components in the original coordinate system are the given values $A_x = 2$, $A_y = 1$, and $A_z = 3$. Substituting these values into (15-4), we obtain

$$\begin{Bmatrix} \bar{A}_x \\ \bar{A}_y \\ \bar{A}_z \end{Bmatrix} = \begin{bmatrix} 0.866 & 0.500 & 0 \\ -0.500 & 0.866 & 0 \\ 0 & 0 & 1 \end{bmatrix} \begin{Bmatrix} 2 \\ 1 \\ 3 \end{Bmatrix} = \begin{Bmatrix} 2.232 \\ -0.134 \\ 3.000 \end{Bmatrix}$$

Do these values seem reasonable from the figure?

Example

The airplane in Figure 15-2 is in a climbing turn, climbing at the angle θ while rolled toward the left at the angle ψ. In order to analyze the motion of the airplane, we want to put the forces acting on it into a single rectangular Cartesian coordinate system. What are the components of lift L, drag D, thrust T, and gravitational force f_g, in the coordinate system with axes in the radial, forward, and vertical directions?

Looking first at the thrust T and aerodynamic forces L and D, we see that these are acting along axes of pitch, roll, and yaw, and that to bring these into coincidence with the desired axes will require two successive rotations. Let us denote by **A** the resultant of thrust and aerodynamic forces, and denote the pitch, roll, and yaw axes as x, y, and z, respectively. We then have

$$A_x = 0 \qquad A_y = T - D \qquad A_z = L$$

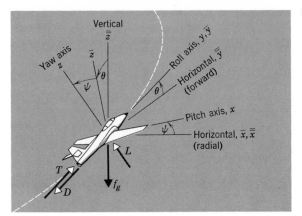

Figure 15-2

The direction cosines of the first rotation (right handed, angle ψ about the roll axis) are

$$l_{\bar{x}x} = \cos \psi \qquad\qquad l_{\bar{x}y} = \cos \frac{\pi}{2} \qquad l_{\bar{x}z} = \cos\left(\frac{\pi}{2} + \psi\right)$$
$$= 0 \qquad\qquad\qquad = -\sin \psi$$

$$l_{\bar{y}x} = \cos \frac{\pi}{2} \qquad\qquad l_{\bar{y}y} = \cos 0 \qquad l_{\bar{y}z} = \cos \frac{\pi}{2}$$
$$= 0 \qquad\qquad\qquad\qquad = 1 \qquad\qquad\qquad = 0$$

$$l_{\bar{z}x} = \cos\left(\frac{\pi}{2} - \psi\right) \qquad l_{\bar{z}y} = \cos \frac{\pi}{2} \qquad l_{\bar{z}x} = \cos \psi$$
$$= \sin \psi \qquad\qquad\qquad = 0$$

Equation 15-4 then gives the $\bar{x}$, $\bar{y}$, and $\bar{z}$ components of the thrust and aerodynamic forces, as

$$\begin{Bmatrix} \bar{A}_x \\ \bar{A}_y \\ \bar{A}_z \end{Bmatrix} = \begin{bmatrix} \cos \psi & 0 & -\sin \psi \\ 0 & 1 & 0 \\ \sin \psi & 0 & \cos \psi \end{bmatrix} \begin{Bmatrix} 0 \\ T - D \\ L \end{Bmatrix} = \begin{Bmatrix} -L \sin \psi \\ T - D \\ L \cos \psi \end{Bmatrix}$$

For the second rotation (left handed, angle θ about the x axis) the direction cosines are

$$l_{\bar{\bar{x}}\bar{x}} = 1 \qquad l_{\bar{\bar{x}}\bar{y}} = 0 \qquad l_{\bar{\bar{x}}\bar{z}} = 0$$
$$l_{\bar{\bar{y}}\bar{x}} = 0 \qquad l_{\bar{\bar{y}}\bar{y}} = \cos \theta \qquad l_{\bar{\bar{y}}\bar{z}} = -\sin \theta$$
$$l_{\bar{\bar{z}}\bar{x}} = 0 \qquad l_{\bar{\bar{z}}\bar{y}} = \sin \theta \qquad l_{\bar{\bar{z}}\bar{z}} = \cos \theta$$

Another application of Equation 15-4, using these values and the above values of $\bar{A}_x$, $\bar{A}_y$, and $\bar{A}_z$, gives us

$$
\begin{Bmatrix} \bar{\bar{A}}_x \\ \bar{\bar{A}}_y \\ \bar{\bar{A}}_z \end{Bmatrix} = \begin{bmatrix} 1 & 0 & 0 \\ 0 & \cos\theta & -\sin\theta \\ 0 & \sin\theta & \cos\theta \end{bmatrix} \begin{Bmatrix} -L\sin\psi \\ T - D \\ L\cos\psi \end{Bmatrix}
$$

$$
= \begin{Bmatrix} -L\sin\psi \\ (T-D)\cos\theta - L\cos\psi\sin\theta \\ (T-D)\sin\theta + L\cos\psi\cos\theta \end{Bmatrix}
$$

The resultant of all forces acting on the airplane is now obtained by adding the gravitational component to those just analyzed:

$$
\mathbf{F} = \mathbf{A} - f_g\bar{\mathbf{u}}_z
$$

$$
\begin{Bmatrix} \bar{\bar{F}}_x \\ \bar{\bar{F}}_y \\ \bar{\bar{F}}_z \end{Bmatrix} = \begin{Bmatrix} -L\sin\psi \\ (T-D)\cos\theta - L\cos\psi\sin\theta \\ (T-D)\sin\theta + L\cos\psi\cos\theta - f_g \end{Bmatrix}
$$

The two transformations that led to the components $\{\bar{\bar{A}}\}$ can be combined in the single equation

$$
\begin{Bmatrix} \bar{\bar{A}}_x \\ \bar{\bar{A}}_y \\ \bar{\bar{A}}_z \end{Bmatrix} = \begin{bmatrix} 1 & 0 & 0 \\ 0 & \cos\theta & -\sin\theta \\ 0 & \sin\theta & \cos\theta \end{bmatrix} \begin{bmatrix} \cos\psi & 0 & -\sin\psi \\ 0 & 1 & 0 \\ \sin\psi & 0 & \cos\psi \end{bmatrix} \begin{Bmatrix} 0 \\ T - D \\ L \end{Bmatrix}
$$

Now, the multiplication of the column and the adjacent square matrix was the first step taken in the above analysis. But since the associative law holds for matrix multiplication, we may also carry out the multiplication of the two square matrices first, with the result

$$
\begin{Bmatrix} \bar{\bar{A}}_x \\ \bar{\bar{A}}_y \\ \bar{\bar{A}}_z \end{Bmatrix} = \begin{bmatrix} \cos\psi & 0 & -\sin\psi \\ -\sin\theta\sin\psi & \cos\theta & -\sin\theta\cos\psi \\ \cos\theta\sin\psi & \sin\theta & \cos\theta\cos\psi \end{bmatrix} \begin{Bmatrix} 0 \\ T - D \\ L \end{Bmatrix}
$$

With this latter choice of multiplication order, we now have a square matrix that gives a direct transformation from the x,y,z components to the $\bar{\bar{x}},\bar{\bar{y}},\bar{\bar{z}}$ components. Carrying out the multiplication indicated in this last equation produces the same result obtained previously.

The Inverse Rotation.   Suppose that, instead of using Equation 15-4 to obtain the $\bar{x},\bar{y},\bar{z}$ components in terms of the x,y,z components, we *begin* with the $\bar{x},\bar{y},\bar{z}$ components. Then the x,y,z components could be obtained by

$$
\begin{Bmatrix} A_x \\ A_y \\ A_z \end{Bmatrix} = \begin{bmatrix} l_{x\bar{x}} & l_{x\bar{y}} & l_{x\bar{z}} \\ l_{y\bar{x}} & l_{y\bar{y}} & l_{y\bar{z}} \\ l_{z\bar{x}} & l_{z\bar{y}} & l_{z\bar{z}} \end{bmatrix} \begin{Bmatrix} \bar{A}_x \\ \bar{A}_y \\ \bar{A}_z \end{Bmatrix}
$$

where $l_{i\bar{j}} = \cos \measuredangle^{\bar{j}}_i$. But since

$$
l_{i\bar{j}} = \cos \measuredangle^{\bar{j}}_i = \cos \measuredangle^{i}_{\bar{j}} = l_{\bar{j}i}
$$

this last equation could also be written as

$$
\begin{Bmatrix} A_x \\ A_y \\ A_z \end{Bmatrix} = \begin{bmatrix} l_{\bar{x}x} & l_{\bar{y}x} & l_{\bar{z}x} \\ l_{\bar{x}y} & l_{\bar{y}y} & l_{\bar{z}y} \\ l_{\bar{x}z} & l_{\bar{y}z} & l_{\bar{z}z} \end{bmatrix} \begin{Bmatrix} \bar{A}_x \\ \bar{A}_y \\ \bar{A}_z \end{Bmatrix}
$$

Now, observe that the square matrix in this last equation is the transpose of that appearing in Equation 15-4; that is, Equation 15-4 and this last equation may be written as

$$
\{\bar{A}\} = [l]\{A\} \tag{a}
$$

$$
\{A\} = [l]^T\{\bar{A}\} \tag{b}
$$

But Equation b can also be obtained from Equation a by premultiplying by the inverse of $[l]$:

$$
\begin{aligned}
[l]^{-1}\{\bar{A}\} &= [l]^{-1}[l]\{A\} \\
&= [1]\{A\} \\
&= \{A\} \tag{c}
\end{aligned}
$$

Comparison of (c) with (b), and account of the fact that these equations hold arbitrary $\{A\}$, lead to the important relationship

$$
[l]^{-1} = [l]^T \tag{15-5}
$$

That is, *the inverse of a matrix representing a rigid rotation of axes is equal to its transpose.* A matrix with this property is said to be *orthonormal.* It is instructive to verify that multiplication of any of the matrices of direction cosines in the preceding examples by its transpose will yield the identity matrix.

Problems

15-12 Referring to Figure 15-2, determine the x-y-z resolution of the resultant force acting on the airplane.

15-13 Multiply the square matrix at the bottom of p. 61 by its transpose and simplify.

15-14 Show that

$$\{\bar{A}\}^T\{\bar{B}\} = \{A\}^T\{B\}$$

where

$$\{\bar{A}\} = [l]\{A\} \qquad \{\bar{B}\} = [l]\{B\}$$

What does this result mean?

15-15 Show that

$$[\bar{A}\times]\{\bar{B}\} = [l][A\times]\{B\}$$

where

$$\{\bar{A}\} = [l]\{A\} \qquad \{\bar{B}\} = [l]\{B\}$$

What does this result mean?

15-16 Refer to Figure 15-2. Starting with the airplane in a level flight orientation (aligned with $\bar{\bar{x}}$-$\bar{\bar{y}}$-$\bar{\bar{z}}$) let it first rotate about the roll axis through the angle ψ, then about the pitch axis through the angle θ. How does the final position compare with that achieved by the sequence indicated in the figure (first θ about the pitch axis, followed by ψ about the roll axis)? Do this with an object such as a toy airplane, letting each angle of rotation be 90°.

15-17 Write out the direction cosine matrix for each of the two sequences of rotations described in the previous problem and compare them.

15-18 Let the pitch and roll angles of the previous problem be small enough that the approximations

$$\cos\phi = 1 - \frac{\phi^2}{2!} + \cdots \simeq 1 \qquad \sin\phi = \phi - \frac{\phi^3}{3!} + \cdots \simeq \phi$$

are valid. How do the rotation matrices for the two different sequences compare in this case?

15-3

DERIVATIVES OF A VECTOR IN MATRIX NOTATION. We have made extensive use of the representation for the derivative of a vector,

$$\overset{\alpha}{\dot{\mathbf{A}}} = \dot{A}_x\mathbf{u}_x + \dot{A}_y\mathbf{u}_y + \dot{A}_z\mathbf{u}_z$$

where we must understand that this is valid only if the unit base vectors remain fixed in the α reference frame. We write the matrix equivalent of this as

$$
{}_\alpha\mathbf{A}:\quad {}_\alpha\left\{\begin{array}{c} \dot{A}_x \\ \dot{A}_y \\ \dot{A}_z \end{array}\right\} \tag{15-6a}
$$

If the orientation of the x,y,z axes is fixed in a reference frame β, which rotates with angular velocity $\boldsymbol{\Omega}$ with respect to α, then the matrix of derivatives of components will represent the β-observed derivative of the vector $\mathbf{A}$,

$$
{}_\beta\mathbf{A}:\quad {}_\beta\left\{\begin{array}{c} \dot{A}_x \\ \dot{A}_y \\ \dot{A}_z \end{array}\right\} \tag{15-6b}
$$

which is different from the α-observed derivative of $\mathbf{A}$. The coordinate axes for the above resolutions of the two vectors $\overset{\alpha}{\mathbf{A}}$ and $\overset{\beta}{\mathbf{A}}$ coincide *only instantaneously*, so that the elements in ${}_\alpha\{\dot{A}\}$ and ${}_\beta\{\dot{A}\}$ differ.

The basic relationship for this situation,

$$
\overset{\alpha}{\mathbf{A}} = \overset{\beta}{\mathbf{A}} + \boldsymbol{\Omega} \times \mathbf{A} \tag{14-1}
$$

can be written in matrix notation with the help of Equation 15-2, as

$$
{}_\alpha\left\{\begin{array}{c} \dot{A}_x \\ \dot{A}_y \\ \dot{A}_z \end{array}\right\} = {}_\beta\left\{\begin{array}{c} \dot{A}_x \\ \dot{A}_y \\ \dot{A}_z \end{array}\right\} + \begin{bmatrix} 0 & -\Omega_z & \Omega_y \\ \Omega_z & 0 & -\Omega_x \\ -\Omega_y & \Omega_x & 0 \end{bmatrix} \left\{\begin{array}{c} A_x \\ A_y \\ A_z \end{array}\right\}
$$

or, in abbreviated form,

$$
\boxed{ {}_\alpha\{\dot{A}\} = {}_\beta\{\dot{A}\} + [\Omega \times]\{A\} } \tag{15-7}
$$

Example

A reference frame β is rotating relative to a reference frame α at a constant 2 rad/s about an axis fixed in both reference frames. Let a set of coordinate axes x,y,z be fixed in the α reference frame and a set of coordinate axes $\bar{x},\bar{y},\bar{z}$ be fixed in the β reference frame. Let the z and $\bar{z}$ axes coincide with the axis of rotation and let the x,y axes coincide with the $\bar{x},\bar{y}$ axes at time $t = 0$. As shown in Figure 15-3, a vector $\mathbf{A}(t)$ varies in a manner such that its $\bar{x}$ component, initially zero, increases at a constant rate of 3 unit/s, and its $\bar{y}$ and $\bar{z}$ components remain a constant 1 unit each. Verify each of the resolutions of the vectors in the following, considering carefully the application of the indicated equation to obtain each set of components from others shown.

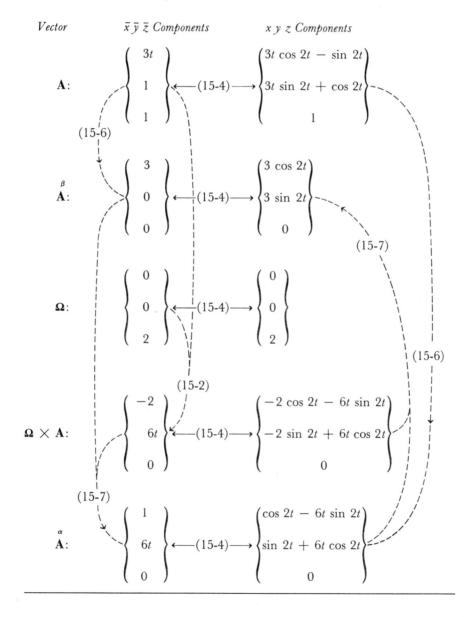

Vector $\bar{x}\,\bar{y}\,\bar{z}$ *Components*   *x y z Components*

$\mathbf{A}$:
$$\begin{Bmatrix} 3t \\ 1 \\ 1 \end{Bmatrix} \longleftarrow(15\text{-}4)\longrightarrow \begin{Bmatrix} 3t\,\cos 2t\,-\,\sin 2t \\ 3t\,\sin 2t\,+\,\cos 2t \\ 1 \end{Bmatrix}$$

(15-6)

$\overset{\beta}{\mathbf{A}}$:
$$\begin{Bmatrix} 3 \\ 0 \\ 0 \end{Bmatrix} \longleftarrow(15\text{-}4)\longrightarrow \begin{Bmatrix} 3\,\cos 2t \\ 3\,\sin 2t \\ 0 \end{Bmatrix}$$

(15-7)

$\mathbf{\Omega}$:
$$\begin{Bmatrix} 0 \\ 0 \\ 2 \end{Bmatrix} \longleftarrow(15\text{-}4)\longrightarrow \begin{Bmatrix} 0 \\ 0 \\ 2 \end{Bmatrix}$$

(15-6)

(15-2)

$\mathbf{\Omega}\times\mathbf{A}$:
$$\begin{Bmatrix} -2 \\ 6t \\ 0 \end{Bmatrix} \longleftarrow(15\text{-}4)\longrightarrow \begin{Bmatrix} -2\,\cos 2t\,-\,6t\,\sin 2t \\ -2\,\sin 2t\,+\,6t\,\cos 2t \\ 0 \end{Bmatrix}$$

(15-7)

$\overset{\alpha}{\mathbf{A}}$:
$$\begin{Bmatrix} 1 \\ 6t \\ 0 \end{Bmatrix} \longleftarrow(15\text{-}4)\longrightarrow \begin{Bmatrix} \cos 2t\,-\,6t\,\sin 2t \\ \sin 2t\,+\,6t\,\cos 2t \\ 0 \end{Bmatrix}$$

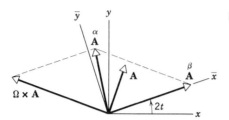

Figure 15-3

In summary: any vector, such as $\mathbf{A}$, $\overset{\alpha}{\mathbf{A}}$, $\overset{\beta}{\mathbf{A}}$, $\mathbf{\Omega}$, and so on, may be resolved onto any set of coordinate axes. The resulting components depend, at that instant, only on the instantaneous *orientation* of the axes, not on their motion. The derivatives of the components in $\{A\}$ form the components of the derivative, *observed in the reference frame to which the coordinate axes are fixed*, of the vector $\mathbf{A}$.

Problems

15-19 Rework Problem 14-1, using matrix notation.

15-20 A set of coordinate axes x,y,z are fixed in the reference frame α and the coordinate axes $\bar{x},\bar{y},\bar{z}$ are fixed in the reference frame β. The coordinate axes coincide at $t = 0$. For the vector $\mathbf{A}(t)$ and constant angular velocity $_\alpha\mathbf{\Omega}_\beta$ given in Problem 14-1, complete a table like that on p. 65.

15-21 Suppose that the maneuver being performed by the airplane of Figure 15-2 is uniform; that is, θ and ψ are constant, and the speed v of the airplane and radius c of the helical path are constant. Determine the x-y-z resolution and the $\bar{x}$-$\bar{y}$-$\bar{z}$ resolution of the vector $\dot{\mathbf{F}}$, where $\mathbf{F}$ is the force discussed in the example and the dot indicates the earth-observed derivative.

15-22 For circular cylindrical coordinates:
(a) Show that the direction cosines giving the $\mathbf{u}_\rho - \mathbf{u}_\phi - \mathbf{u}_z$ resolution of a vector in terms of the $\mathbf{u}_x - \mathbf{u}_y - \mathbf{u}_z$ resolution are the elements in the matrix

$$[\phi] = \begin{bmatrix} \cos\phi & \sin\phi & 0 \\ -\sin\phi & \cos\phi & 0 \\ 0 & 0 & 1 \end{bmatrix}$$

(b) What are the $\mathbf{u}_\rho = \mathbf{u}_\phi - \mathbf{u}_z$ components of the position vector $\mathbf{r}$?
(c) The $\mathbf{u}_x - \mathbf{u}_y - \mathbf{u}_z$ resolutions of velocity and acceleration are

$$
\left\{
\begin{array}{c}
v_x \\
v_y \\
v_z
\end{array}
\right\}
=
\left\{
\begin{array}{c}
\dot{x} \\
\dot{y} \\
\dot{z}
\end{array}
\right\}
\qquad
\left\{
\begin{array}{c}
a_x \\
a_y \\
a_z
\end{array}
\right\}
=
\left\{
\begin{array}{c}
\ddot{x} \\
\ddot{y} \\
\ddot{z}
\end{array}
\right\}
$$

The $\mathbf{u}_x - \mathbf{u}_y - \mathbf{u}_z$ resolutions of these vectors are then

$$
\left\{
\begin{array}{c}
v_\rho \\
v_\phi \\
v_z
\end{array}
\right\}
= [\phi]
\left\{
\begin{array}{c}
\dot{x} \\
\dot{y} \\
\dot{z}
\end{array}
\right\}
= [\phi] \frac{d}{dt} [\phi]^{-1}
\left\{
\begin{array}{c}
\rho \\
0 \\
z
\end{array}
\right\}
$$

$$
\left\{
\begin{array}{c}
a_\rho \\
a_\phi \\
a_z
\end{array}
\right\}
= [\phi]
\left\{
\begin{array}{c}
\ddot{x} \\
\ddot{y} \\
\ddot{z}
\end{array}
\right\}
= [\phi] \frac{d^2}{dt^2} [\phi]^{-1}
\left\{
\begin{array}{c}
\rho \\
0 \\
z
\end{array}
\right\}
$$

Carry out in detail the multiplication and differentiation indicated here and verify the result with Equation 10-7. In terms of effort and opportunity for errors, how does this approach compare with that of Problem 14-9?

15-23   For the spherical coordinates shown in Figure 10-8:
(a) Write the matrix of direction cosines giving the $\mathbf{u}_r$-$\mathbf{u}_\phi$-$\mathbf{u}_\theta$ resolution of a vector in terms of the $\mathbf{u}_\rho$-$\mathbf{u}_\phi$-$\mathbf{u}_z$ resolution. With this result and that of the previous problem, show that the matrix

$$
\begin{bmatrix}
\cos\phi\cos\theta & \sin\phi\cos\theta & \sin\theta \\
-\sin\phi & \cos\phi & 0 \\
-\cos\phi\sin\theta & -\sin\phi\sin\theta & \cos\theta
\end{bmatrix}
$$

gives the $\mathbf{u}_r$-$\mathbf{u}_\phi$-$\mathbf{u}_\theta$ resolution of a vector in terms of the $\mathbf{u}_x$-$\mathbf{u}_y$-$\mathbf{u}_z$ resolution.
(b) Starting with the $\mathbf{u}_r$-$\mathbf{u}_\phi$-$\mathbf{u}_\theta$ resolution of the position vector, use the result from (a) above to verify Equations 10-8.

15-4

LINEAR VECTOR TRANSFORMATIONS. Transformations that are somewhat more general than $\{\bar{A}\} = [l]\{A\}$ arise in a number of applications.

For example, the transformation from rectangular Cartesian coordinates x to oblique Cartesian coordinates ξ, turns out to have the form

$$\{A_\xi\} = [S]\{A_x\}$$

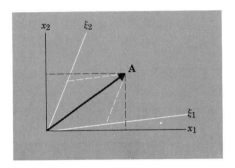

in which the elements in $[S]$ are not restricted by the orthogonality relationships (15-5). Another example in which the mathematical structure is the same is the relationship (6-6),* which gives the moment resulting from a linear distribution of pressure on a flat surface, in terms of the unit vector **s**, along the zero-pressure line.

$$\begin{Bmatrix} M_x \\ M_y \end{Bmatrix} = \begin{bmatrix} kI_{xx} & kI_{xy} \\ kI_{yx} & kI_{yy} \end{bmatrix} \begin{Bmatrix} s_x \\ s_y \end{Bmatrix} \qquad [6\text{-}6]$$

An example closer to our immediate interest is the relationship between the angular velocity and angular momentum of a rigid body. Equations 13-1 give the form of this for the case in which the z axis is aligned with the angular velocity vector:

$$\begin{Bmatrix} H_x \\ H_y \\ H_z \end{Bmatrix} = \begin{bmatrix} \text{---} & \text{---} & I_{xz} \\ \text{---} & \text{---} & I_{yz} \\ \text{---} & \text{---} & I_{zz} \end{bmatrix} \begin{Bmatrix} 0 \\ 0 \\ \omega \end{Bmatrix} \qquad [13\text{-}1]$$

You can undoubtedly guess the form for a more general alignment of coordinate axes, which will be derived in Chapter 16. Similar relationships between pairs of physically related vectors arise in the analysis of stress, of strain, of heat conduction in anisotropic media, of the bending of beams, and in many other situations.

All of the above examples have the form

$$\{B\} = [T]\{A\} \qquad (15\text{-}8)$$

---

* Charles Smith, *Statics*, Wiley, 1976, pp. 138–139.

In the case of the transformation from one Cartesian coordinate system to another, we interpret $\{A\}$ and $\{B\}$ as the sets of components of the same vector in two different coordinate systems. In the remaining examples, we interpret $\{A\}$ and $\{B\}$ as the sets of components of related but different vectors, both **A** and **B** being referred to the same coordinate system.

Components of a Vector Operator in Different Coordinate Systems. When Equation 15-8 relates two physical entities represented by the vectors **A** and **B**, the elements in the matrices $\{A\}$ and $\{B\}$ depend on the coordinate system employed. The elements in $[T]$ must similarly depend on the coordinate system; for, if a change in coordinates is made, so that $\{A_x\}$ becomes $\{A_\xi\}$ $(= [S]\{A_x\})$, then a new set of values $[T_\xi]$ will be required so that the relationship (15-8) correctly gives $\{B_\xi\}$ $(= [S]\{B_x\})$. The point is important enough to warrant repetition in terms of a specific example.

The angular velocity and angular momentum of a spinning rigid body are vectors, having no dependence on the particular coordinate system we happen to select to represent them. Similarly, the distribution of the mass of the body, which determines the **ω-H** relationship, is a physical entity that has no such dependence. It is only the *representation* of the inertial properties of the body, in the matrix of second moments of mass, which depends on the coordinate system. This idea will now be pursued to determine the specific law for transforming the elements of a square matrix that represents such a vector operator.

In the x coordinate system, the **A-B** relationship appears as

$$\{B_x\} = [T_x]\{A_x\} \tag{a}$$

while in the ξ coordinate system the same relationship appears as

$$\{B_\xi\} = [T_\xi]\{A_\xi\} \tag{b}$$

But since **A** and **B** are *vectors*, their components in the two coordinate systems are related by

$$\{B_x\} = [S]^{-1}\{B_\xi\} \qquad \{A_x\} = [S]^{-1}\{A_\xi\}$$

If these are substituted into (a), and each member premultiplied by $[S]$, there results

$$\{B_\xi\} = [S][T_x][S]^{-1}\{A_\xi\} \tag{c}$$

This will agree with (b) if we have*

$$\boxed{[T_\xi] = [S][T_x][S]^{-1}} \tag{15-9}$$

---

* The nine equations (15-9) are certainly *sufficient* to satisfy the three equations resulting from combining (a), (b), and (c). They are also *necessary* if we demand that the relationships (b) and (c) hold for arbitrary $\{A_\xi\}$.

If the coordinate transformation is orthonormal, $[S] = [l]$; in this case, because of Equation 15-5, Equation 15-9 reduces to

$$[\bar{T}] = [l][T][l]^T \qquad (15\text{-}10)$$

The elements in the square matrices $[T_\xi]$ and $[T_x]$, which satisfy Equation 15-9, are called the Cartesian components of a *tensor* of second order. In the same terminology, a vector is called a tensor of first order, and a scalar a tensor of zeroth order.

Example

The moment vector for a submerged plate is given in the x,y coordinates by

$$\left\{ \begin{array}{c} M_x \\ M_y \end{array} \right\} = k \begin{bmatrix} I_{xx} & I_{xy} \\ I_{yx} & I_{yy} \end{bmatrix} \left\{ \begin{array}{c} s_x \\ s_y \end{array} \right\}$$

Determine the components $k[\bar{I}]$ that represent the **s-M** operator in the $\bar{x},\bar{y}$ coordinates, which are obtained from the x,y coordinates by a counterclockwise rotation through the angle θ.

The transformation of coordinates is given by

$$\left\{ \begin{array}{c} \bar{x} \\ \bar{y} \end{array} \right\} = \begin{bmatrix} \cos\theta & \sin\theta \\ -\sin\theta & \cos\theta \end{bmatrix} \left\{ \begin{array}{c} x \\ y \end{array} \right\}$$

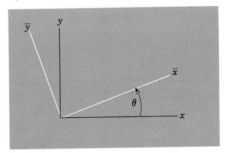

The matrix $[l]$ thus identified, we substitute into Equation 15-10:

$$k\begin{bmatrix} \bar{I}_{xx} & \bar{I}_{xy} \\ \bar{I}_{yx} & \bar{I}_{yy} \end{bmatrix} = k \begin{bmatrix} \cos\theta & \sin\theta \\ -\sin\theta & \cos\theta \end{bmatrix} \begin{bmatrix} I_{xx} & I_{xy} \\ I_{yx} & I_{yy} \end{bmatrix} \begin{bmatrix} \cos\theta & -\sin\theta \\ \sin\theta & \cos\theta \end{bmatrix}$$

$$= k \begin{bmatrix} \begin{array}{c} I_{xx}\cos^2\theta + I_{yy}\sin^2\theta \\ + 2I_{xy}\cos\theta\sin\theta \end{array} & \begin{array}{c} (I_{yy} - I_{xx})\cos\theta\sin\theta \\ + I_{xy}(\cos^2\theta - \sin^2\theta) \end{array} \\ \begin{array}{c} (I_{yy} - I_{xx})\cos\theta\sin\theta \\ + I_{xy}(\cos^2\theta - \sin^2\theta) \end{array} & \begin{array}{c} I_{xx}\sin^2\theta + I_{yy}\cos^2\theta \\ - 2I_{xy}\cos\theta\sin\theta \end{array} \end{bmatrix}$$

Observe that this agrees with Equations* 6-9.

---

* Charles Smith, *Statics*, Wiley, 1976, p. 139.

Principal Directions.   In every application for a vector relationship of the form

$$\{B\} = [T]\{A\}$$

the following question has important significance. Given $[T]$, for what value(s) of $\{A\}$ will the related vector $\{B\} = [T]\{A\}$ lie in the same direction as $\{A\}$?

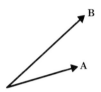

When $[T]$ represents the inertia tensor for a rigid body, this is equivalent to asking, "For what axis of spin will the angular momentum and angular velocity coincide?"

The vector with this special property is called an *eigenvector* of the operator represented by $[T]$; its direction is referred to as a *principal direction*. To find such a direction from the given matrix $[T]$, observe first that **A** and **B** will be parallel if and only if $\mathbf{B} = \lambda\mathbf{A}$. Thus, we write

$$\{B\} = [T]\{A\} = \lambda\{A\} \qquad (15\text{-}11a)$$

Written out in full, the equation $[T]\{A\} - \lambda\{A\} = \{0\}$ has the form

$$\begin{bmatrix} T_{xx} - \lambda & T_{xy} & T_{xz} \\ T_{yx} & T_{yy} - \lambda & T_{yz} \\ T_{zx} & T_{zy} & T_{zz} - \lambda \end{bmatrix} \begin{Bmatrix} A_x \\ A_y \\ A_z \end{Bmatrix} = \begin{Bmatrix} 0 \\ 0 \\ 0 \end{Bmatrix} \qquad (15\text{-}11b)$$

An obvious trivial solution, $A_1 = A_2 = A_3 = 0$, is of little interest. A theorem of algebra* states that a nontrivial solution exists if and only if the determinant of the above square matrix vanishes:

$$\begin{vmatrix} T_{xx} - \lambda & T_{xy} & T_{xz} \\ T_{yx} & T_{yy} - \lambda & T_{yz} \\ T_{zx} & T_{zy} & T_{zz} - \lambda \end{vmatrix} = 0 \qquad (15\text{-}12a)$$

Upon expansion, this becomes

$$\lambda^3 - J_1\lambda^2 + J_2\lambda - J_3 = 0 \qquad (15\text{-}12b)$$

* See, for instance, G. Hadley, *Linear Algebra*, Addison-Wesley, 1961.

where

$$J_1 = T_{xx} + T_{yy} + T_{zz}$$

$$J_2 = \begin{vmatrix} T_{yy} & T_{yz} \\ T_{zy} & T_{zz} \end{vmatrix} + \begin{vmatrix} T_{xx} & T_{xz} \\ T_{zx} & T_{zz} \end{vmatrix} + \begin{vmatrix} T_{xx} & T_{xy} \\ T_{yx} & T_{yy} \end{vmatrix}$$

$$J_3 = \begin{vmatrix} T_{xx} & T_{xy} & T_{xz} \\ T_{yx} & T_{yy} & T_{yz} \\ T_{zx} & T_{zy} & T_{zz} \end{vmatrix} \tag{15-13}$$

The roots of Equation 15-12, λ_1, λ_2, and λ_3, are called the *eigenvalues* of $[T]$. They may be found by widely published methods for solving cubic equations.

In proceeding further to find the corresponding eigenvectors, several fundamental facts concerning this problem should be kept in mind. Proofs of the theorems may be found in the many books concerned with linear algebra and will not be reproduced here.

1. Because (for all our applications) the coefficients J_1, J_2, and J_3 are all real, at least one of the three roots of Equation 5-12 is real; if one root is complex, another is its complex conjugate. Hence, we have either one or three real roots in every case.

2. When the matrix $[T]$ is *symmetric* (i.e., $[T]^T = [T]$, as in the case of the matrix representing the inertia tensor of a rigid body), all three roots are real.

3. To each of the three eigenvalues corresponds an eigenvector, determined by Equation 15-11 with the eigenvalue substituted therein. When one of the eigenvalues is so substituted, no more than two of the three equations (15-11) are independent; this means that the direction of the corresponding eigenvector may be determined, but its magnitude is arbitrary.

4. When the matrix $[T]$ is symmetric, the eigenvectors are mutually perpendicular.

Once an eigenvalue is known, the corresponding eigenvector may be found from Equation 15-11. In practice it is convenient to assign a value of 1 to one of the components of $\{A\}$ (the magnitude of $\mathbf{A}$ is arbitrary) and use two of the three equations to find the other two components.

As an example, let us find the eigenvalues and corresponding eigenvectors of the matrix

$$\begin{bmatrix} 3 & 2 & 1 \\ 2 & 2 & 1 \\ 1 & 1 & 1 \end{bmatrix}$$

Equation 15-12 becomes, for this case,

$$\begin{vmatrix} (3-\lambda) & 2 & 1 \\ 2 & (2-\lambda) & 1 \\ 1 & 1 & (1-\lambda) \end{vmatrix} = 0$$

Or,

$$\lambda^3 - 6\lambda^2 + 5\lambda - 1 = 0$$

The roots of this equation are

$$\lambda_1 = 0.3080$$
$$\lambda_2 = 0.6431$$
$$\lambda_3 = 5.0489$$

Substituting λ_1 into Equation 15-11 yields

$$\begin{bmatrix} 2.6920 & 2.0000 & 1.0000 \\ 2.0000 & 1.6920 & 1.0000 \\ 1.0000 & 1.0000 & 0.6920 \end{bmatrix} \begin{Bmatrix} A_x \\ A_y \\ A_z \end{Bmatrix} = \begin{Bmatrix} 0 \\ 0 \\ 0 \end{Bmatrix}$$

If we let $A_x = 1$, the first two equations may be written as

$$2.0000\,A_y + A_z = -2.6920$$
$$1.6920\,A_y + A_z = -2.0000$$

These two equations may be solved for A_y and A_z, with the result

$$A_y = -2.2470 \qquad A_z = 1.8020$$

Thus, an eigenvector corresponding to λ_1 is

$$\begin{Bmatrix} A_x \\ A_y \\ A_z \end{Bmatrix} = \begin{Bmatrix} 1.0000 \\ -2.2470 \\ 1.8020 \end{Bmatrix}$$

Substituting λ_2 into Equation 15-11b, and repeating the procedure just outlined, results in an eigenvector corresponding to this eigenvalue:

$$\begin{Bmatrix} A_x \\ A_y \\ A_z \end{Bmatrix}_2 = \begin{Bmatrix} 1.0000 \\ -0.5550 \\ -1.2470 \end{Bmatrix}$$

The third principal direction may be determined in the same manner. Following is a summary of the results:

$$\lambda_1 = 0.3080 \qquad\qquad \lambda_2 = 0.6431 \qquad\qquad \lambda_3 = 5.0489$$

$$\{A\}_1 = \begin{Bmatrix} 1.0000 \\ -2.2470 \\ 1.8020 \end{Bmatrix} \qquad \{A\}_2 = \begin{Bmatrix} 1.0000 \\ -0.5550 \\ -1.2470 \end{Bmatrix} \qquad \{A\}_3 = \begin{Bmatrix} 1.0000 \\ 0.8019 \\ 0.4450 \end{Bmatrix}$$

You may find it instructive to premultiply each of these eigenvectors by the original matrix $[T]$. You may also wish to check to see that they are mutually perpendicular.

Problems

15-24 Two vectors **A** and **B** are related by

$$\begin{Bmatrix} B_x \\ B_y \end{Bmatrix} = \begin{bmatrix} 3 & 1 \\ 1 & 3 \end{bmatrix} \begin{Bmatrix} A_x \\ A_y \end{Bmatrix}$$

For each of the different indicated values of **A**, evaluate and sketch the corresponding value of **B**.

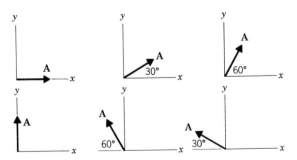

Try to guess directions for which **A** and **B** coincide. What is the effect on **B** of a change in magnitude of **A**?

15-25 Repeat Problem 15-24, where now **A** and **B** are related by

$$\begin{Bmatrix} B_x \\ B_y \end{Bmatrix} = \begin{bmatrix} \dfrac{\sqrt{3}}{2} & \dfrac{1}{2} \\ -\dfrac{1}{2} & \dfrac{\sqrt{3}}{2} \end{bmatrix} \begin{Bmatrix} A_x \\ A_y \end{Bmatrix}$$

15-26 Use Equation 15-10 to determine the representation of the operator of Problem 15-24 in a coordinate system that is rotated 45° counterclockwise from the x,y axes. Rework Problem 15-24 for the same set of values of **A**. (Note that the same **A** will now have a different set of components.)

15-27 Use Equation 15-11 and the procedures that follow to determine the principal directions for the operator in Problem 15-24.

15-28 Determine the principal direction(s) for the operator represented by

$$\begin{bmatrix} \dfrac{\sqrt{3}}{2} & \dfrac{1}{2} & 0 \\ -\dfrac{1}{2} & \dfrac{\sqrt{3}}{2} & 0 \\ 0 & 0 & 1 \end{bmatrix}$$

15-29 Determine the eigenvalues and corresponding eigenvectors of the following matrices:

(a) $\begin{bmatrix} 2 & 0 & 3 \\ 0 & 4 & 1 \\ 3 & 1 & 1 \end{bmatrix}$ (b) $\begin{bmatrix} 6 & 1 & 1 \\ 1 & -1 & 2 \\ 1 & 2 & 0 \end{bmatrix}$

(c) $\begin{bmatrix} -1 & 2 & -1 \\ 2 & -4 & 2 \\ -1 & 2 & -1 \end{bmatrix}$ (d) $\begin{bmatrix} 2 & 0 & 0 \\ 0 & 3 & 0 \\ 0 & 0 & -1 \end{bmatrix}$

15-30 For the example of pp. 73–74:
(a) Verify that the eigenvectors are mutually perpendicular.
(b) Write expressions for unit vectors in the principal directions.
(c) Use the unit vectors to form a matrix of direction cosines that will transform the original x,y,z coordinates to a set of axes aligned with the principal directions.

(d) For this transformation, apply Equation 15-10.

15-31   In analyzing the state of stress at a point, consideration of the equilibrium of the element leads to

$$
\begin{Bmatrix} S_x \\ S_y \end{Bmatrix} = \begin{bmatrix} \sigma_{xx} & \sigma_{yx} \\ \sigma_{xy} & \sigma_{yy} \end{bmatrix} \begin{Bmatrix} n_x \\ n_y \end{Bmatrix}
$$

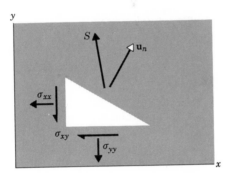

where **S** is the resultant force per unit area on the surface to which $\mathbf{u}_n$ is normal, and the σ_{ij} are components of force per unit area on the coordinate planes. For given values of σ_{xx}, $\sigma_{xy} = \sigma_{yx}$, and σ_{yy}, consider various directions given by different $\mathbf{u}_n$. For what values of $\mathbf{u}_n$ will **S** be normal to the inclined surface?

15-32   Given a symmetric matrix $[T]$ we wish to find a transformation of coordinates $[l]$ that will diagonalize $[T]$, that is, such that $[\overline{T}]$, given by

$$
[\overline{T}] = [l][T][l]^T
$$

has zeros in all elements except those in the main diagonal. Show that each row in $[l]$ is made up of the components of an eigenvector of $[T]$, that is, satisfies Equation 15-11. *Suggestion.* Premultiply Equation 15-10 by $[l]^T$ and equate columns in the result.

15-33   Show that the analog of Equation 15-7, for a tensor represented in rectangular Cartesian coordinates by $[T]$, is

$$
_\alpha[\dot{T}] = {}_\beta[\dot{T}] + [\Omega\times][T] - [T][\Omega\times]
$$

and that if $[T]$ is symmetric, this can also be written as

$$
_\alpha[\dot{T}] = {}_\beta[\dot{T}] + [\Omega\times][T] + [[\Omega\times][T]]^T
$$

Suggestion. Note that the two vectors **A** and **B** in the relationship $\{B\} = [T]\{A\}$ obey Equation 15-7.

RIGID
BODY
DYNAMICS

In this chapter we examine the procedures for analyzing the motions of a rigid body in general. As anyone who has played with a toy gyroscope or performed the experiment suggested at the beginning of Chapter 13 realizes, the dynamic behavior of rigid bodies can be most fascinating. From a practical viewpoint, behavior that surprises the designer of a system with rotating components can be very expensive.

16-1

DISPLACEMENT KINEMATICS. Although less information will often fulfill practical requirements, a *complete* analysis of the mechanical behavior of a rigid body will involve knowledge of the variation of its position and orientation with time. Specification of its position and orientation in a reference frame requires six quantities; three components of a position vector may be introduced to specify the location of a selected point on the body; three other quantities are then required to specify its "angular" position, or orientation.

A proper selection of these latter three quantities must be done with care. To begin with, let us note that the tempting concept of an angular displacement

Figure 16-1

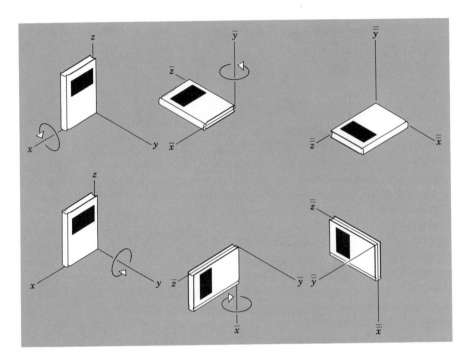

vector, in analogy with the successfully employed angular velocity vector, is unsatisfactory. The chief difficulty is revealed by the simple experiment illustrated in Figure 16-1. Here are shown the results of two successive rotations of 90° each about the x and y axes; in one case a rotation about the x axis is followed by a rotation about the $\bar{y}$ axis, and in the other case the order is reversed. The difference between the final positions in the two cases is most striking.* In addition to this marked dependence on the order of the rotations, observe that in neither case can the final result be achieved by a single "resultant" rotation about an axis lying in the plane of the two axes involved in the "component" rotations. Thus if we introduce a vector lying along the axis of rotation to specify each of the "components" of rotation, vector addition will fail to give the resultant of two rotations about the different axes. In fact, it is not at all obvious whether it is *possible* to achieve the final position by a single equivalent rotation. The answer is provided in the following theorem:

---

* A further experiment will point up a more striking result: in each sequence let the axis to be used for the second rotation be fixed, rather than carried with the book during the first rotation; compare the result with that of Figure 16-1.

Euler's Theorem on Rigid Rotations. The most general change in configuration of a rigid body with one point fixed may be achieved by a single rotation about a fixed axis. To demonstrate this, consider the three noncollinear points attached to the body, as shown in Figure 16.2. Let point O be the fixed point and let points P and Q move to the positions P' and Q' respectively. The positions of these points completely determine the positions of all points in the body, so that a rotation that will simultaneously carry P to P' and Q to Q' will be the rotation we seek.

Referring to Figure 16-2a, we note first that if point P is to reach point P' via a rotation about an axis through O, this axis must lie in the plane that is perpendicular to and bisects the line PP'. An equivalent statement is true of an axis about which Q may be rotated to reach Q'. Therefore, if the same line is to serve as an axis of rotation for carrying both P to P' and Q to Q', it is necessary that it lie in both planes. Figure 16-2b shows the case in which these two planes do not coincide. We select a point C lying on the line of intersection of these two planes, and draw lines from points P, P', Q, and Q' to C. Now from the fact that the body is rigid, triangles $P'OQ'$ and POQ are congruent, and from symmetry the lengths $\overline{CP'}$ and $\overline{CQ'}$ are respectively equal to lengths $\overline{CP}$ and $\overline{CQ}$. It then follows that moving the tetrahedron $COPQ$ rigidly until P and Q coincide respectively with P' and Q', leaves point C in its original position. The change in position of the rigid body can therefore be achieved by a rotation about the fixed axis OC. For the case in which the two planes bisecting PP' and QQ' coincide, choose point C to lie on the line of intersection of OPQ and $OP'Q'$, and repeat the above argument.

Can you now locate the axes for single equivalent rotations of the book in Figure 16-1?

An analogous theorem for two dimensions states that the most general plane rigid displacement can be achieved by a rotation about the fixed axis

Figure 16-2

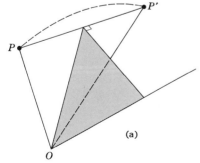

(a)

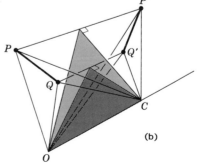

(b)

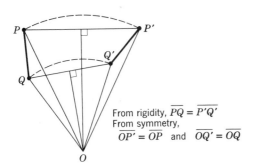

Figure 16-3

From rigidity, $\overline{PQ} = \overline{P'Q'}$
From symmetry,
$\overline{OP'} = \overline{OP}$ and $\overline{OQ'} = \overline{OQ}$

perpendicular to the plane of the displacement. Figure 16-3 suggests a proof, and a means of determining the location of the axis of rotation in terms of the initial and final positions of two point P and Q.

Chasles' Theorem.   The most general change in configuration of a rigid body can be achieved by a translation and a rotation about a fixed axis. The statement is almost self-evident once we have Euler's theorem. Let all points on the body shift parallel to the line connecting the initial and final positions of an arbitrarily selected point. Then, by Euler's theorem, a rotation about a fixed axis running through this point will bring the body to the final configuration. The direction and magnitude of the translation will depend on the point selected since, in general, the lines connecting initial and final positions of different points are neither parallel nor of the same length. However, the results developed in the next paragraph will reveal that the orientation of the axis of rotation and the amount of rotation are independent of the point selected.

Screw Displacement.   An informative resolution of the general displacement of a rigid body is developed as follows. Let the displacement be achieved by a translation that moves point P to its final position P', followed by a rotation about a fixed axis running through P'. But instead of accomplishing the translation along the straight line PP', move the body first parallel to the axis of subsequent rotation, bringing point P to point p, then perpendicular to this axis until p reaches P', as shown in Figure 16-4. But during the second stage of displacement, and the subsequent rotation, the movement of each point on the body is in a *plane* perpendicular to the axis of rotation; as can be seen from Figure 16-3, the same result may be achieved by a rotation about a parallel axis. The planar part of the displacement has no net effect on the position of points on this latter axis. The line connecting their initial and final positions is called the *screw axis*. Now, the same total displacement could be achieved by a simultaneous translation parallel to and rotation about the screw axis. Such a path is aptly termed a *screw displacement*. The most general planar displacement

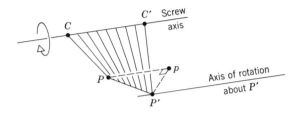

Figure 16-4

and the most general spatial displacement with one point fixed are the special case in which the pitch of the screw is zero.

The student who hasn't already done so should notice the analogy between the above developments and those in Chapter 14 concerning instantaneous *motions* of a rigid body. All but one of the conclusions reached there, concerning the various ways of resolving such motions, can be verified from the facts we have just presented, by viewing the motions as being made up of small increments of displacement which occur during small increments of time. The exception is the vector character of angular velocity, which, as we have seen, angular displacements do not possess. We will see presently that *small increments* of angular displacement *can* be correctly described by vectors, making the connection between the kinematics of rigid body displacements and rigid body motions complete.

Matrix Analysis for Rotational Displacements. *Orthogonal transformations* are so termed because they carry three mutually perpendicular axes to a new orientation in which they are still mutually perpendicular. Because the matrix $[l]$ in Equation 15-4 has this property, we will find the analysis there to be useful in the study of angular displacements. The procedure for analyzing the orientation of the rigid body in a reference frame will be based on specification of the direction cosines between a set of axes fixed in the body and another set fixed in the reference frame.

The orthonormality property of the matrix $[l]$ is reflected in a set of restrictions that its elements must satisfy. Written out in detail, the equations implied by $[l]^T[l] = [1]$ are

$$l_{\bar{x}x}l_{\bar{y}x} + l_{\bar{x}y}l_{\bar{y}y} + l_{\bar{x}z}l_{\bar{y}z} = 0$$
$$l_{\bar{y}x}l_{\bar{z}x} + l_{\bar{y}y}l_{\bar{z}y} + l_{\bar{y}z}l_{\bar{z}z} = 0 \qquad (16\text{-}1a)$$
$$l_{\bar{z}x}l_{\bar{x}x} + l_{\bar{z}y}l_{\bar{x}y} + l_{\bar{z}z}l_{\bar{x}z} = 0$$

$$l_{\bar{x}x}^2 + l_{\bar{x}y}^2 + l_{\bar{x}z}^2 = 1$$
$$l_{\bar{y}x}^2 + l_{\bar{y}y}^2 + l_{\bar{y}z}^2 = 1 \qquad (16\text{-}1b)$$
$$l_{\bar{z}x}^2 + l_{\bar{z}y}^2 + l_{\bar{z}z}^2 = 1$$

Thus, there is considerable interdependence among the nine direction cosines. It will follow from a later result—Equations 16-4 and 16-5—that specification of the four quantities $(l_{\bar{x}x} + l_{\bar{y}y} + l_{\bar{z}z})$, $(l_{\bar{x}y} - l_{\bar{y}x})$, $(l_{\bar{y}z} - l_{\bar{z}x})$, and $(l_{\bar{z}x} - l_{\bar{x}z})$ will completely specify each of the nine elements in the matrix $[l]$.

Special cases that are often of interest are rotations about the coordinate axes themselves. It is instructive to verify that the transformation matrices correctly describe the accompanying figures, and that the orthogonality relationships are satisfied:

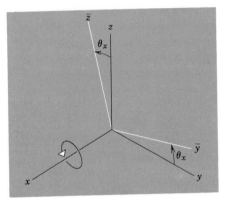

$$\left\{ \begin{matrix} \bar{x} \\ \bar{y} \\ \bar{z} \end{matrix} \right\} = \begin{bmatrix} 1 & 0 & 0 \\ 0 & \cos\theta_x & \sin\theta_x \\ 0 & -\sin\theta_x & \cos\theta_x \end{bmatrix} \left\{ \begin{matrix} x \\ y \\ z \end{matrix} \right\} \qquad (16\text{-}2a)$$

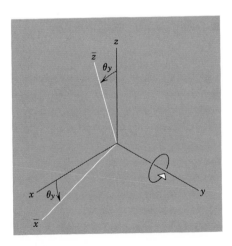

$$\left\{ \begin{array}{c} \bar{x} \\ \bar{y} \\ \bar{z} \end{array} \right\} = \begin{bmatrix} \cos\theta_y & 0 & -\sin\theta_y \\ 0 & 1 & 0 \\ \sin\theta_y & 0 & \cos\theta_y \end{bmatrix} \left\{ \begin{array}{c} x \\ y \\ z \end{array} \right\} \qquad (16\text{-}2b)$$

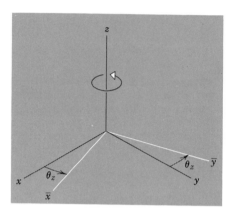

$$\left\{ \begin{array}{c} \bar{x} \\ \bar{y} \\ \bar{z} \end{array} \right\} = \begin{bmatrix} \cos\theta_z & \sin\theta_z & 0 \\ -\sin\theta_z & \cos\theta_z & 0 \\ 0 & 0 & 1 \end{bmatrix} \left\{ \begin{array}{c} x \\ y \\ z \end{array} \right\} \qquad (16\text{-}2c)$$

Direction Cosines and the Single, Fixed-Axis Rotation.   The orthogonal matrices that we have examined in some detail are not the only means—indeed, in many situations far from the most useful means—for specifying a rigid rotation. Another possibility, suggested by Euler's theorem, is the specification of the orientation of the axis about which the rotation can be achieved and the angle of rotation. The quantities in this system are, of course, related to the direction cosines we have been analyzing, and we now examine the relationships between these two systems.

Suppose the rotation is known in terms of the orientation of, and the angle of rotation about, the axis of Euler's theorem. We wish to determine the direction cosine matrix that specifies the same rotation. That is, in terms of the direction cosines of the axis R and the angle Φ, shown in Figure 16-5a, we wish to express the elements of the rotation operator that relates the $\bar{x},\bar{y},\bar{z}$ and the x,y,z axes.

Now, if the *same* rotation is referred to the ξ, η, ζ coordinates of Figure 16-5b, with ξ selected along the rotation axis R, the matrix representing the

Figure 16-5

Direction of rotation for positive Φ is defined by the right-hand rule associated with the R axis.

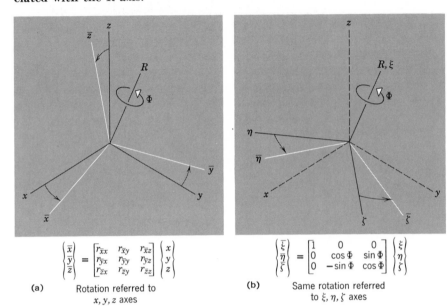

$$\begin{Bmatrix} \bar{x} \\ \bar{y} \\ \bar{z} \end{Bmatrix} = \begin{bmatrix} r_{\bar{x}x} & r_{\bar{x}y} & r_{\bar{x}z} \\ r_{\bar{y}x} & r_{\bar{y}y} & r_{\bar{y}z} \\ r_{\bar{z}x} & r_{\bar{z}y} & r_{\bar{z}z} \end{bmatrix} \begin{Bmatrix} x \\ y \\ z \end{Bmatrix}$$

(a)   Rotation referred to
              x, y, z axes

$$\begin{Bmatrix} \bar{\xi} \\ \bar{\eta} \\ \bar{\zeta} \end{Bmatrix} = \begin{bmatrix} 1 & 0 & 0 \\ 0 & \cos\Phi & \sin\Phi \\ 0 & -\sin\Phi & \cos\Phi \end{bmatrix} \begin{Bmatrix} \xi \\ \eta \\ \zeta \end{Bmatrix}$$

(b)   Same rotation referred
              to ξ, η, ζ axes

rotation takes the form accompanying this figure. Denoting the matrix of the *coordinate transformation* relating the x,y,z and ξ,η,ζ systems by $[l]$,

$$\begin{Bmatrix} x \\ y \\ z \end{Bmatrix} = \begin{bmatrix} l_{x\xi} & l_{x\eta} & l_{x\zeta} \\ l_{y\xi} & l_{y\eta} & l_{y\zeta} \\ l_{z\xi} & l_{z\eta} & l_{z\zeta} \end{bmatrix} \begin{Bmatrix} \xi \\ \eta \\ \zeta \end{Bmatrix}$$

we note that since we selected the ξ axis to coincide with R

$$l_{x\xi} = R_x$$
$$l_{y\xi} = R_y$$
$$l_{z\xi} = R_z$$

Referring to Equation 15-10, and noting that for the present case the *operator T* is represented in the ξ,η,ζ coordinates by the matrix accompanying Figure 16-5*b*, we write the x,y,z representation of this same operator by

$$
\begin{bmatrix} r_{\bar{x}x} & r_{\bar{x}y} & r_{\bar{x}z} \\ r_{\bar{y}x} & r_{\bar{y}y} & r_{\bar{y}z} \\ r_{\bar{z}x} & r_{\bar{z}y} & r_{\bar{z}z} \end{bmatrix} = \begin{bmatrix} R_x & l_{x\eta} & l_{x\zeta} \\ R_y & l_{y\eta} & l_{y\zeta} \\ R_z & l_{z\eta} & l_{z\zeta} \end{bmatrix} \begin{bmatrix} 1 & 0 & 0 \\ 0 & \cos \Phi & \sin \Phi \\ 0 & -\sin \Phi & \cos \Phi \end{bmatrix} \begin{bmatrix} R_x & R_y & R_z \\ l_{x\eta} & l_{y\eta} & l_{z\eta} \\ l_{x\zeta} & l_{y\zeta} & l_{z\zeta} \end{bmatrix}
$$

The multiplication on the right-hand side yields

$$
\begin{aligned}
r_{\bar{x}x} &= R_x^2 + (l_{x\eta}^2 + l_{x\zeta}^2) \cos \Phi \\
r_{\bar{x}y} &= R_x R_y + (l_{x\eta} l_{y\eta} + l_{x\zeta} l_{y\zeta}) \cos \Phi + (l_{x\eta} l_{y\zeta} - l_{y\eta} l_{x\zeta}) \sin \Phi \\
r_{\bar{x}z} &= R_x R_z + (l_{x\eta} l_{z\eta} + l_{x\zeta} l_{z\zeta}) \cos \Phi - (l_{z\eta} l_{x\zeta} - l_{x\eta} l_{z\zeta}) \sin \Phi \\
r_{\bar{y}x} &= R_y R_x + (l_{y\eta} l_{x\eta} + l_{y\zeta} l_{x\zeta}) \cos \Phi - (l_{x\eta} l_{y\zeta} - l_{y\eta} l_{x\zeta}) \sin \Phi \\
r_{\bar{y}y} &= R_y^2 + (l_{y\eta}^2 + l_{y\zeta}^2) \cos \Phi \\
r_{\bar{y}z} &= R_y R_z + (l_{y\eta} l_{z\eta} + l_{y\zeta} l_{z\zeta}) \cos \Phi + (l_{y\eta} l_{z\zeta} - l_{z\eta} l_{y\zeta}) \sin \Phi \\
r_{\bar{z}x} &= R_z R_x + (l_{z\eta} l_{x\eta} + l_{z\zeta} l_{x\zeta}) \cos \Phi + (l_{z\eta} l_{x\zeta} - l_{x\eta} l_{z\zeta}) \sin \Phi \\
r_{\bar{z}y} &= R_z R_y + (l_{z\eta} l_{y\eta} + l_{z\zeta} l_{y\zeta}) \cos \Phi - (l_{y\eta} l_{z\zeta} - l_{z\eta} l_{y\zeta}) \sin \Phi \\
r_{\bar{z}z} &= R_z^2 + (l_{z\eta}^2 + l_{z\zeta}^2) \cos \Phi
\end{aligned}
$$

Since the final result cannot be expected to depend on the orientation of the η and ζ axes, it should be no surprise that the ls may be eliminated from these expressions. This can be accomplished by means of the orthonormality relationships (16-1), and the following, which can be derived from the vector relationship $\mathbf{u}_x = \mathbf{u}_y \times \mathbf{u}_z$:

$$
\begin{aligned}
R_x &= l_{y\eta} l_{z\zeta} - l_{z\eta} l_{y\zeta} \\
R_y &= l_{z\eta} l_{x\zeta} - l_{x\eta} l_{z\zeta} \\
R_z &= l_{x\eta} l_{y\zeta} - l_{y\eta} l_{x\zeta}
\end{aligned}
$$

These orthonormality equations permit reduction of the above to

$$
\begin{bmatrix} r_{\bar{x}x} & r_{\bar{x}y} & r_{\bar{x}z} \\ r_{\bar{y}x} & r_{\bar{y}y} & r_{\bar{y}z} \\ r_{\bar{z}x} & r_{\bar{z}y} & r_{\bar{z}z} \end{bmatrix} = \cos \Phi \begin{bmatrix} 1 & 0 & 0 \\ 0 & 1 & 0 \\ 0 & 0 & 1 \end{bmatrix} + (1 - \cos \Phi) \begin{bmatrix} R_x^2 & R_x R_y & R_x R_z \\ R_y R_x & R_y^2 & R_y R_z \\ R_z R_x & R_z R_y & R_z^2 \end{bmatrix}
$$

$$
+ \sin \Phi \begin{bmatrix} 0 & R_z & -R_y \\ -R_z & 0 & R_x \\ R_y & -R_x & 0 \end{bmatrix} \tag{16-3}
$$

giving us the direction cosines in terms of the orientation (R_x, R_y, R_z) of the axis of rotation, and the angle Φ of rotation.

The inverse problem is easily solved from Equation 16-3. Given the direction cosines r_{ij}, the orientation of the rotation axis and the angle of rotation may be determined as follows. Noting that $R_x^2 + R_y^2 + R_z^2 = 1$, we have, from Equation 16-3,

$$r_{\bar{x}x} + r_{\bar{y}y} + r_{\bar{z}z} = \cos \Phi(1 + 1 + 1) + (1 - \cos \Phi)1$$

from which we may determine Φ:

$$\cos \Phi = \tfrac{1}{2}(r_{\bar{x}x} + r_{\bar{y}y} + r_{\bar{z}z} - 1) \tag{16-4}$$

Next, observe from Equation 16-3, that

$$\tfrac{1}{2}(r_{\bar{y}z} - r_{\bar{z}y}) = R_x \sin \Phi$$
$$\tfrac{1}{2}(r_{\bar{z}x} - r_{\bar{x}z}) = R_y \sin \Phi \tag{16-5}$$
$$\tfrac{1}{2}(r_{\bar{x}y} - r_{\bar{y}x}) = R_z \sin \Phi$$

from which R_x, R_y, and R_z can then be computed.*

Observe that the positive sense for Φ was chosen according to a right-hand rotation around the R axis, carrying the x,y,z axes into the $\bar{x},\bar{y},\bar{z}$ axes. The components of a fixed vector $\mathbf{A}$, in the two coordinate systems, are related by

$$\{\bar{A}\} = [r]\{A\}$$

Now, the inverse, or transpose, of $[r]$ will effect the same rotation (i.e., right handed, through angle Φ) *of a vector* in a *fixed* set of coordinate axes. The same conclusion may be reached from Equation 16-3: note first that replacement of Φ by $-\Phi$ will effect a left-handed rotation of the coordinate axes or, equivalently, a right-handed rotation of a vector in a fixed set of axes. But since $\cos(-\Phi) = \cos \Phi$ and $\sin(-\Phi) = -\sin \Phi$, it follows that replacement of Φ with $-\Phi$ is also equivalent to transposing the right-hand side of Equation 16-3.

As an example, let us find the orientation of the axis about which a single rotation will achieve the equivalent of the two successive rotations of the book illustrated in the upper sequence of Figure 16-1. Refer the rotations to a set of fixed x,y,z axes. Axes fixed in the book and originally coincident with these will be denoted as $\bar{x},\bar{y},\bar{z}$, after the first rotation, and as $\bar{\bar{x}},\bar{\bar{y}},\bar{\bar{z}}$, after the second rotation. The first rotation is given by

* In the special case where $\Phi = \pi$, these equations will fail to specify R. In this case,

$$R_i = \cos\left(\frac{1}{2}\cos^{-1} r_{\bar{i}i}\right) = \pm\sqrt{\frac{1 + r_{\bar{i}i}}{2}} \qquad i = x,y,z$$

$$\left\{\begin{array}{c}\bar{x}\\[6pt]\bar{y}\\[6pt]\bar{z}\end{array}\right\} = \begin{bmatrix} 1 & 0 & 0 \\[6pt] 0 & \cos\dfrac{\pi}{2} & \sin\dfrac{\pi}{2} \\[6pt] 0 & -\sin\dfrac{\pi}{2} & \cos\dfrac{\pi}{2} \end{bmatrix} \left\{\begin{array}{c}x\\[6pt]y\\[6pt]z\end{array}\right\}$$

and the second by

$$\left\{\begin{array}{c}\bar{\bar{x}}\\[6pt]\bar{\bar{y}}\\[6pt]\bar{\bar{z}}\end{array}\right\} = \begin{bmatrix} \cos\dfrac{\pi}{2} & 0 & -\sin\dfrac{\pi}{2} \\[6pt] 0 & 1 & 0 \\[6pt] \sin\dfrac{\pi}{2} & 0 & \cos\dfrac{\pi}{2} \end{bmatrix} \left\{\begin{array}{c}\bar{x}\\[6pt]\bar{y}\\[6pt]\bar{z}\end{array}\right\}$$

Combining the two equations gives us the direction cosines between the x,y,z and $\bar{\bar{x}},\bar{\bar{y}},\bar{\bar{z}}$ axes,

$$\{\bar{\bar{x}}\} = [r]\{x\}$$

where

$$[r] = \begin{bmatrix} 0 & 0 & -1 \\ 0 & 1 & 0 \\ 1 & 0 & 0 \end{bmatrix}\begin{bmatrix} 1 & 0 & 0 \\ 0 & 0 & 1 \\ 0 & -1 & 0 \end{bmatrix} = \begin{bmatrix} 0 & 1 & 0 \\ 0 & 0 & 1 \\ 1 & 0 & 0 \end{bmatrix}$$

The angle of rotation is given by Equation 16-4:

$$\cos\Phi = \tfrac{1}{2}(0 + 0 + 0 - 1) = -\tfrac{1}{2}$$

from which*

$$\Phi = \frac{2\pi}{3}$$

The direction cosines of the axis of rotation are given by Equation 16-5:

$$R_1 \sin\Phi = \tfrac{1}{2}[1 - 0] = \tfrac{1}{2}$$
$$R_2 \sin\Phi = \tfrac{1}{2}[1 - 0] = \tfrac{1}{2}$$
$$R_3 \sin\Phi = \tfrac{1}{2}[1 - 0] = \tfrac{1}{2}$$

* The other choice, $\Phi = -(2\pi/3)$, leads to the negative of the R axis given by $\Phi = +(2\pi/3)$; with the sense of the rotation determined by the right-hand rule, the result is identical.

With the above value of Φ these are

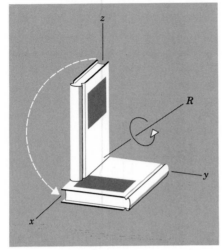

$$R_1 = \frac{1}{\sqrt{3}}$$

$$R_2 = \frac{1}{\sqrt{3}}$$

$$R_3 = \frac{1}{\sqrt{3}}$$

Infinitesimal Rotations.   Here we examine those rotations of which the angle is small enough so that the approximations

$$\sin \Phi = \Phi - \frac{\Phi^2}{3!} + \cdots \simeq \Phi$$

$$\cos \Phi = 1 - \frac{\Phi^2}{2!} + \cdots \simeq 1$$

are valid. These approximations reduce the somewhat complicated algebra of finite rotations, to congruity with the simpler vector operations that we have used to analyze angular velocities. In particular, vector *addition* will correctly give the resultant of two or more infinitesimal rotations. This is in agreement with Equation 14-9, p. 36, from which we view a simultaneous spinning about two or more axes as equivalent to a spinning about an axis in the direction of the vector sum of the several angular velocity components.

Let us verify these statements. The above approximations reduce the rotation matrix (16-3) to

$$[r] = [1] + [\epsilon]$$

where, with the small angle of rotation denoted by $\delta\Phi$,

$$[\epsilon] = \delta\Phi \begin{bmatrix} 0 & R_z & -R_y \\ -R_z & 0 & R_x \\ R_y & -R_x & 0 \end{bmatrix}$$

Now consider a vector **A**, fixed in a rigid body that undergoes a small rotation. The change in the vector, $\delta\mathbf{A}$, will have the components in the column

$$
\begin{aligned}
\{\delta A\} &= [r]^T\{A\} - \{A\} \\
&= [[r]^T - [1]]\{A\} \\
&= [\epsilon]^T\{A\}
\end{aligned}
$$

Or, written in full,

$$
\begin{Bmatrix} \delta A_x \\ \delta A_y \\ \delta A_z \end{Bmatrix} = \begin{bmatrix} 0 & -R_z\delta\Phi & R_y\delta\Phi \\ R_z\delta\Phi & 0 & -R_x\delta\Phi \\ -R_y\delta\Phi & R_x\delta\Phi & 0 \end{bmatrix} \begin{Bmatrix} A_x \\ A_y \\ A_z \end{Bmatrix}
$$

A comparison with Equation 15-2 reveals that this represents the vector relationship

$$
\delta\mathbf{A} = \delta\boldsymbol{\Phi} \times \mathbf{A}
$$

where $\delta\boldsymbol{\Phi}$ is a vector having magnitude equal to the small angle of rotation, and direction along the axis of rotation R. This is identical with the change we

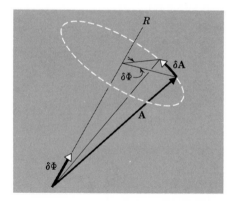

would observe in the vector **A** fixed in a reference frame spinning with angular velocity $\boldsymbol{\Omega}$, during a small time increment δt:

$$
\delta\mathbf{A} = \boldsymbol{\Omega}\delta t \times \mathbf{A}
$$

Finally, to establish that a succession of such small rotations may be correctly represented by the sum of the vectors of the small rotations, consider a small rotation $[\epsilon_1]$, followed by a small rotation $[\epsilon_2]$. In this case, the changes in the vector components are given by

$$
\begin{aligned}
\{\delta A\} &= [[1] + [\epsilon_2]^T][[1] + [\epsilon_1]^T]\{A\} - \{A\} \\
&= [[1] + [\epsilon_1]^T + [\epsilon_2]^T + [\epsilon_2]^T[\epsilon_1]^T]\{A\} - \{A\} \\
&= [[\epsilon_1]^T + [\epsilon_2]^T + [\epsilon_2]^T[\epsilon_1]^T]\{A\} \\
&\simeq [[\epsilon_1]^T + [\epsilon_2]^T]\{A\} \\
&= [\epsilon_1]^T\{A\} + [\epsilon_2]^T\{A\}
\end{aligned}
$$

Or, written as vector operations, this is equivalent to

$$\delta \mathbf{A} \simeq (\delta \boldsymbol{\Phi}_1 + \delta \boldsymbol{\Phi}_2) \times \mathbf{A} = \delta \boldsymbol{\Phi}_1 \times \mathbf{A} + \delta \boldsymbol{\Phi}_2 \times \mathbf{A}$$

Thus, the vectors that represent *small* angular displacements may be added in the ordinary way to give the vector representing the resultant of the rotations. The effect of the order in which the rotations are performed is expressed in the negligible term

$$[\epsilon_2]^T[\epsilon_1]^T\{A\}$$

As an example, consider a rigid body that is rotated through 5° about the z axis, then 9° about the body-fixed x axis (both rotations right handed). First, let us determine the change in a unit vector attached to the body, originally aligned with the y axis. The projections of the twice-rotated unit vector onto the original axes will be given by

$$
\left[
\begin{bmatrix}
1 & 0 & 0 \\
0 & \cos 9° & \sin 9° \\
0 & -\sin 9° & \cos 9°
\end{bmatrix}
\begin{bmatrix}
\cos 5° & \sin 5° & 0 \\
-\sin 5° & \cos 5° & 0 \\
0 & 0 & 1
\end{bmatrix}
\right]^T
\begin{Bmatrix} 0 \\ 1 \\ 0 \end{Bmatrix}
=
\begin{Bmatrix} -0.086\ 09 \\ 0.983\ 93 \\ 0.155\ 83 \end{Bmatrix}
$$

The change in the vector then has the components

$$
\delta \mathbf{A}:\
\begin{Bmatrix} -0.086\ 09 \\ 0.983\ 93 \\ 0.155\ 83 \end{Bmatrix}
-
\begin{Bmatrix} 0 \\ 1 \\ 0 \end{Bmatrix}
=
\begin{Bmatrix} -0.0861 \\ -0.0161 \\ 0.1558 \end{Bmatrix}
$$

Now let us compare this exact result with the change computed according to $\delta \mathbf{A} = \delta \boldsymbol{\Phi}_1 \times \mathbf{A} + \delta \boldsymbol{\Phi}_2 \times \mathbf{A}$. Referred to the x,y,z coordinates, the rotation vectors have the components

$$\delta\Phi_1: \frac{5\pi}{180} \left\{ \begin{array}{c} 0 \\ 0 \\ 1 \end{array} \right\} = \left\{ \begin{array}{cc} 0 & \\ 0 & \\ 0.087 & 27 \end{array} \right\}$$

$$\delta\Phi_2: \frac{9\pi}{180} \left\{ \begin{array}{c} \cos 5° \\ \sin 5° \\ 0 \end{array} \right\} = \left\{ \begin{array}{cc} 0.156 & 48 \\ 0.013 & 69 \\ 0 & \end{array} \right\}$$

and the vector **A** has the components

$$\mathbf{A}: \left\{ \begin{array}{c} 0 \\ 1 \\ 0 \end{array} \right\}$$

Carrying out, in terms of these components, the cross multiplication and addition indicated above results in

$$\delta\mathbf{A}: \left\{ \begin{array}{c} -0.0873 \\ 0.0000 \\ 0.1565 \end{array} \right\}$$

The approximation is thus a fairly good one in this case. Problems 16-16 through 16-18 provide additional reinforcement of your understanding of infinitesimal rotations.

Euler's Angles.   These three angles constitute still another means of specifying the orientation of a rigid body in a reference frame. Dynamical equations of rigid body motion usually take a more tractable form in terms of these quantities than in terms of direction cosines or the orientation of and rotation about the rotation axis.

As shown in Figure 16-6, the angles ϕ, χ, and ψ specify the orientation of the x,y,z axes relative to the X,Y,Z axes. The three rotations, which carry the x,y,z axes from coincidence with the X,Y,Z axes to the position shown, take place as follows:

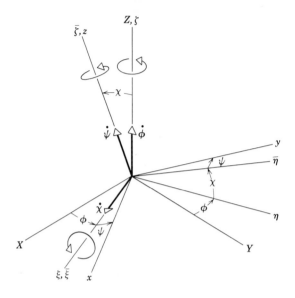

Figure 16-6

1. A rotation through the angle ϕ about the Z axis carries the ξ,η,ζ axes from coincidence with the X,Y,Z axes to the position shown. The matrix of direction cosines of this rotation is

$$
\begin{bmatrix}
\cos\phi & \sin\phi & 0 \\
-\sin\phi & \cos\phi & 0 \\
0 & 0 & 1
\end{bmatrix}
$$

2. A rotation through the angle χ about the ξ axis carries the $\bar{\xi},\bar{\eta},\bar{\zeta}$ axes from coincidence with the ξ,η,ζ axes, to the position shown. The matrix of direction cosines of this rotation is

$$
\begin{bmatrix}
1 & 0 & 0 \\
0 & \cos\chi & \sin\chi \\
0 & -\sin\chi & \cos\chi
\end{bmatrix}
$$

3. A rotation through the angle ψ about the $\bar{\zeta}$ axis carries the x,y,z axes from coincidence with the $\bar{\xi},\bar{\eta},\bar{\zeta}$ axes, to the position shown. The matrix of direction cosines of this rotation is

$$\begin{bmatrix} \cos \psi & \sin \psi & 0 \\ -\sin \psi & \cos \psi & 0 \\ 0 & 0 & 1 \end{bmatrix}$$

With the above matrices, we are in a position to determine the projections of a vector onto any rectangular set of axes in terms of the projections onto any other set.

For example, let us consider the angular velocity $\boldsymbol{\omega}$ of the x,y,z axes relative to the X,Y,Z axes. Referring again to Figure 16-6 we see that there is a component of magnitude $\dot{\phi}$ directed along the Z axis, a component of magnitude $\dot{\chi}$ directed along the ξ axis, and a component of magnitude $\dot{\psi}$ directed along z axis. These components constitute an *oblique* resolution of the angular velocity; that is, the three are not mutually perpendicular. The *rectangular xyz* resolution may be obtained by applying the $[\chi]$ rotation, then the $[\psi]$ rotation to the $\dot{\phi}$ component, the $[\psi]$ rotation to the $\dot{\chi}$ component, then adding these to the $\dot{\psi}$ component:

$$\begin{Bmatrix} \omega_x \\ \omega_y \\ \omega_z \end{Bmatrix} = \begin{bmatrix} \cos \psi & \sin \psi & 0 \\ -\sin \psi & \cos \psi & 0 \\ 0 & 0 & 1 \end{bmatrix} \begin{bmatrix} 1 & 0 & 0 \\ 0 & \cos \chi & \sin \chi \\ 0 & -\sin \chi & \cos \chi \end{bmatrix} \begin{Bmatrix} 0 \\ 0 \\ \dot{\phi} \end{Bmatrix}$$

$$+ \begin{bmatrix} \cos \psi & \sin \psi & 0 \\ -\sin \psi & \cos \psi & 0 \\ 0 & 0 & 1 \end{bmatrix} \begin{Bmatrix} \dot{\chi} \\ 0 \\ 0 \end{Bmatrix} + \begin{Bmatrix} 0 \\ 0 \\ \dot{\psi} \end{Bmatrix}$$

$$= \begin{Bmatrix} \dot{\phi} \sin \chi \sin \psi + \dot{\chi} \cos \psi \\ \dot{\phi} \sin \chi \cos \psi - \dot{\chi} \sin \psi \\ \dot{\phi} \cos \chi + \dot{\psi} \end{Bmatrix} \tag{16-6a}$$

It is instructive to verify that the (X,Y,Z) resolution of this vector is

$$\begin{Bmatrix} \omega_X \\ \omega_Y \\ \omega_Z \end{Bmatrix} = \begin{Bmatrix} \dot{\chi} \cos \phi + \dot{\psi} \sin \phi \sin \chi \\ \dot{\chi} \sin \phi - \dot{\psi} \cos \phi \sin \chi \\ \dot{\phi} + \dot{\psi} \cos \chi \end{Bmatrix} \tag{16-6b}$$

These results, produced in the course of illustrating the techniques for handling the direction cosine matrices related to the Eulerian angles, will prove useful in the analysis of the motions of rigid bodies.

Problems

16-1   Following the method of proving Euler's theorem, p. 79, locate the axis for a single rotation equivalent to each of the sequences of rotations of Figure 16-1.

16-2   Complete the proof of Euler's theorem on rigid rotations, by considering the case where the planes bisecting PP' and QQ' coincide.

16-3   Show that the orientation of the rotation axis and the amount of rotation do not depend on the point selected for the translational part of Chasles' theorem.

16-4   Use Equation 16-3 to derive Equations 16-2a, b, c.

16-5   (a) Verify the statement of the footnote, p. 86.
(b) Verify the statement of the footnote, p. 87.

16-6   Let the angles of rotation about the x and y axes of Figure 16-1 be ϕ and θ, respectively, rather than $\pi/2$. For each sequence of rotations, determine the matrix of direction cosines giving the final orientation of the book relative to the initial orientation.

16-7   Now let us analyze orientations of the book similar to those of Problem 16-6, but where the first rotation of angle ϕ about the x axis (of angle θ about the y axis in the case of the second sequence) is followed by a rotation of angle θ about the y axis (of angle ϕ about the x axis), rather than about the $\bar{y}$ axis (about the $\bar{x}$ axis). For each sequence, determine the matrix of direction cosines giving $\bar{\bar{x}}, \bar{\bar{y}}, \bar{\bar{z}}$ in terms of x,y,z. (*Suggestion.* Use Equation 16-3.) Compare the results with those of Problem 16-6.

16-8   Following is an instructive demonstration of how the resultant of rotations about two different intersecting axes may be achieved by a single equivalent rotation. On a spherical surface attached to the body and with center at the fixed point, we mark a great circle connecting the points A and B, where the rotation axes intersect the sphere, as shown in the diagram. We mark two more great circles passing through point A, making angles of $\alpha/2$ on either side of the first great circle at A. Like circles emanating from point B with angles $\beta/2$ complete a pair of similar spherical triangles ABC and ABC'. Now, a rotation of angle α about OA will bring C to the original position of C', and a subsequent rotation of angle β about OB will return C to its original position. Hence the final result could be achieved equally well by a single rotation about OC.

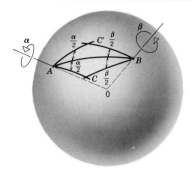

(a) What will be the axis of single equivalent rotation if the order of rotations is reversed?

(b) What will be the axis of single equivalent rotation if the axis for the second rotation, OB, is carried with the body during the first rotation about OA?

(c) For the sequences of questions (a) and (b) above, locate the final position of the point on the sphere originally at B.

(d) What can we now say in general about the order of two successive rotations, about "body-fixed" and "space-fixed" axes?

16-9   Rework Problem 16-7, using the conclusion reached in Problem 16-8, rather than Equation 16-3.

16-10 Evaluate the single, fixed-axis rotation equivalent to lower sequence in Figure 16-1.

16-11 The arm of the radial-arm saw is rotated about Z through 30°, and the motor is rotated about y through 45°. What will be the orientation of a

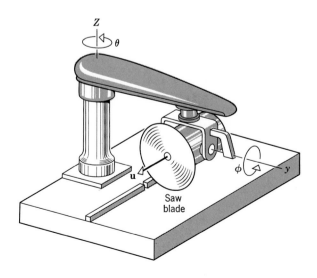

line normal to the saw cut? Evaluate the single, fixed-axis rotation that will effect the same repositioning of the motor.

16-12 Two holes are to be drilled in the part as shown in the sketch. After extracting the drill from hole A, the part is to be repositioned so that the same drill will then cut hole B. The reorientation is to be accomplished by a single rotation about a fixed axis. Locate such an axis and determine the angle of rotation.

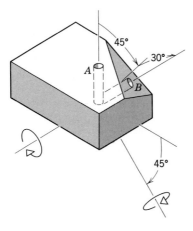

16-13 A part similar to that of Problem 16-12 is to be drilled as indicated. Locate an axis for repositioning by means of a single rotation of the part, and determine the angle of rotation.

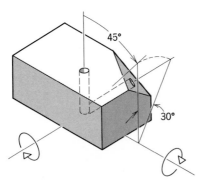

16-14 A part similar to that of Problem 16-12 is to be drilled as indicated, each hole to be normal to the surface on which it is started. Locate an axis for repositioning by means of a single rotation of the part, and determine the angle of rotation.

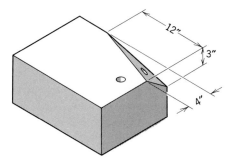

16-15 The antenna, its original line of sight vertical, is given a rotation of 30° about the *A-A* gimbal axis, then a rotation of 45° about the *B-B* gimbal axis. How many degrees from the vertical is its final line of sight?

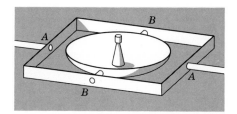

16-16 Rework the example of pp. 90–91, where the 9° rotation is made about the space-fixed *x* axis, rather than the body-fixed *x* axis. Compare the results.

16-17 For the example of pp. 90–91, determine accurately the orientation of the fixed axis of a single equivalent rotation. Compare the result with the vector sum $\delta\Phi_1 + \delta\Phi_2$.

16-18 Repeat the comparison of pp. 90–91 and the previous two problems, where, instead of 5° and 9°, the angles are 15° and 27°.

16-19 For the Euler angle scheme shown, determine the angular velocity of the *x,y,z* triad relative to the *X,Y,Z* triad. Give the projections of this vector on the *x,y,z* axes and on the *X,Y,Z* axes.

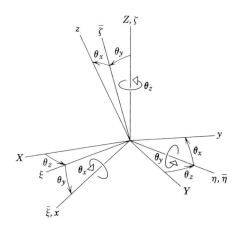

16-20 We want to study the rotations, about its mass center, of a satellite that is orbiting the earth in a circular path of radius R_0. For this purpose, x-y-z axes are fixed to the body and the orientation defined by the angles ψ, θ, ϕ shown. A is normal to the plane of the orbit. Write the x-y-z components of $\boldsymbol{\omega}$, the angular velocity of the body relative to "fixed stars," in terms of the angles shown.

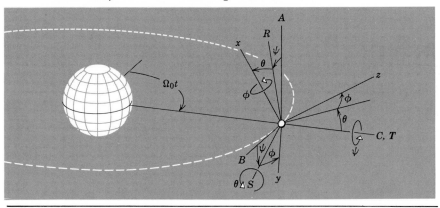

16-2

ANGULAR MOMENTUM AND KINETIC ENERGY. The definition of angular momentum is the same as for the system of n particles (Equation 12-11), but since we now are considering a continuous distribution of mass, the summation is carried out as an integration over the body:

$$\mathbf{H}_A = \int_m \mathbf{r}_{P/A} \times \mathbf{v}_{P/A} \, dm$$

Here P indicates the point at which the element of mass dm is located. If we let point A be attached to the body, the motion of every particle relative to A is completely determined by the angular velocity $\boldsymbol{\omega}$. The angular momentum can then be written directly in terms of $\boldsymbol{\omega}$, by inserting the relationship

$$\mathbf{v}_{P/A} = \boldsymbol{\omega} \times \mathbf{r}_{P/A}$$

into the above integral. Without carrying the subscripts through the remainder of the development, it is to be understood that the point A is the reference for position $\mathbf{r}$, entering the definition of moment of momentum:

$$\mathbf{H} = \int_m \mathbf{r} \times (\boldsymbol{\omega} \times \mathbf{r})dm$$

Further development is facilitated with the introduction of a rectangular Cartesian coordinate system with the origin placed at point A. Insertion of the coordinate expansion for angular velocity,

$$\boldsymbol{\omega} = \omega_x \mathbf{u}_x + \omega_y \mathbf{u}_y + \omega_z \mathbf{u}_z$$

into the above integral leads to

$$\mathbf{H} = \omega_x \int_m \mathbf{r} \times (\mathbf{u}_x \times \mathbf{r})dm + \omega_y \int_m \mathbf{r} \times (\mathbf{u}_y \times \mathbf{r})dm$$
$$+ \omega_z \int_m \mathbf{r} \times (\mathbf{u}_z \times \mathbf{r})dm$$

Observe that the integrals in this expression depend solely on the mass distribution of the body and the placement of the coordinate axes *in the body*. Thus, the magnitude and orientation of $\boldsymbol{\omega}$ *in the body* determine the magnitude and orientation of $\mathbf{H}$ *in the body*, independent of the orientation of the body in a reference frame.

Introducing the abbreviation

$$\mathbf{I}_j = \int_m \mathbf{r} \times (\mathbf{u}_j \times \mathbf{r})dm$$

into the above expansion,

$$\mathbf{H} = \omega_x \mathbf{I}_x + \omega_y \mathbf{I}_y + \omega_z \mathbf{I}_z$$

we see that the integral $\mathbf{I}_j$ may be interpreted as the angular momentum of the body per unit angular velocity in the $\mathbf{u}_j$ direction.

A typical coordinate component of angular momentum, H_i, may be obtained by applying the operation $H_i = \mathbf{u}_i \cdot \mathbf{H}$ to the expansion. Carrying this out for $i = x,y,z$, we have

$$H_x = (\mathbf{u}_x \cdot \mathbf{I}_x)\omega_x + (\mathbf{u}_x \cdot \mathbf{I}_y)\omega_y + (\mathbf{u}_x \cdot \mathbf{I}_z)\omega_z$$
$$H_y = (\mathbf{u}_y \cdot \mathbf{I}_x)\omega_x + (\mathbf{u}_y \cdot \mathbf{I}_y)\omega_y + (\mathbf{u}_y \cdot \mathbf{I}_z)\omega_z \qquad (16\text{-}7)$$
$$H_z = (\mathbf{u}_z \cdot \mathbf{I}_x)\omega_x + (\mathbf{u}_z \cdot \mathbf{I}_y)\omega_y + (\mathbf{u}_z \cdot \mathbf{I}_z)\omega_z$$

The Inertia Tensor. The coefficients of the angular velocity components in Equations 16-7 are called products of inertia, a typical member of the set denoted by I_{ij}:

$$I_{ij} = \mathbf{u}_i \cdot \mathbf{I}_j \qquad (i,j = x,y,z)$$

To give a clearer idea of how these are evaluated, we first apply the identity (3-13) to the integral $\mathbf{I}_j$:

$$I_{ij} = \mathbf{u}_i \cdot \int_m \mathbf{r} \times (\mathbf{u}_j \times \mathbf{r})dm$$

$$= \mathbf{u}_i \cdot \int_m [(\mathbf{r} \cdot \mathbf{r})\mathbf{u}_j - (\mathbf{u}_j \cdot \mathbf{r})\mathbf{r}]dm$$

$$= \mathbf{u}_i \cdot \mathbf{u}_j \int_m \mathbf{r} \cdot \mathbf{r} \, dm - \int_m (\mathbf{u}_i \cdot \mathbf{r})(\mathbf{u}_j \cdot \mathbf{r})dm$$

The results are of two types, depending on whether the subscripts i and j are equal or unequal. Typical of equal subscripts is

$$I_{xx} = \mathbf{u}_x \cdot \mathbf{u}_x \int_m \mathbf{r} \cdot \mathbf{r} \, dm - \int_m (\mathbf{u}_x \cdot \mathbf{r})^2 \, dm$$

$$= \int_m (x^2 + y^2 + z^2)dm - \int_m x^2 \, dm$$

$$= \int_m (y^2 + z^2)dm$$

This is referred to as the *moment of inertia about the x axis*. Similarly, the moments of inertia about the other two axes are

$$I_{yy} = \int_m (z^2 + x^2)dm \qquad I_{zz} = \int_m (x^2 + y^2)dm$$

Typical of unequal subscripts is

$$I_{xy} = \mathbf{u}_x \cdot \mathbf{u}_y \int_m \mathbf{r} \cdot \mathbf{r} \, dm - \int_m (\mathbf{u}_x \cdot \mathbf{r})(\mathbf{u}_y \cdot \mathbf{r})dm$$

$$= -\int_m x \, y \, dm = I_{yx}$$

Similarly,

$$I_{yz} = I_{zy} = -\int_m y \, z \, dm \qquad I_{zx} = I_{xz} = -\int_m z \, x \, dm$$

If one of the coordinate planes is a plane of symmetry, then the products of inertia having one subscript corresponding to the axis normal to the symmetry plane and one subscript corresponding to an axis lying in this plane, will be zero.

This follows from the fact that for every element on one side of the symmetry plane there is a contribution from a corresponding element on the opposite side which, because of opposite sign, cancels the contribution from the first. Although this symmetry is sufficient for the vanishing of these products of inertia, it is not necessary.

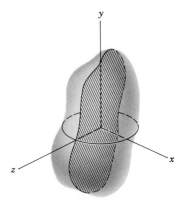

Symmetry about y-z plane

$$I_{xy} = I_{yx} = I_{zx} = I_{xz} = 0$$

Evaluation of the products of inertia is a basic step in evaluating the angular momentum for any rigid body. For homogeneous bodies of fairly simple shapes such as cubes, spheres, circular cones and cylinders, and the like, evaluation is fairly simple. Appendix D lists the results of the integration for some bodies having simple shapes. Numerical approximation is in order where the body has a very complicated shape.

Once the products of inertia have been computed, the angular momentum vector may be determined from Equations 16-7 or their matrix equivalent:

$$
\begin{Bmatrix} H_x \\ H_y \\ H_z \end{Bmatrix} = \begin{bmatrix} I_{xx} & I_{xy} & I_{xz} \\ I_{yx} & I_{yy} & I_{yz} \\ I_{zx} & I_{zy} & I_{zz} \end{bmatrix} \begin{Bmatrix} \omega_x \\ \omega_y \\ \omega_z \end{Bmatrix}
\tag{16-8}
$$

The matrix $[I]$ functions in this relationship as a linear vector operator, assigning a value to the vector $\mathbf{H}$ for any given value of $\boldsymbol{\omega}$. The entity represented in the xyz coordinate system by this matrix is called the *inertia tensor*. It is analogous to the quantity m, which assigns a value to the linear momentum vector $\mathbf{p}$ for any given value of the linear velocity vector $\mathbf{v}$. However, the situation here is considerably more complicated in that, unlike $\mathbf{v}$ and $\mathbf{p}$, $\boldsymbol{\omega}$ and $\mathbf{H}$ do not generally lie in the same direction.

Parallel Shift of Coordinate Axes. The products of inertia, with respect to a set of axes, must often be determined when the values of a parallel set are known. Such a shift of axes is equivalent to transferring the reference point A (the origin of coordinates) about which angular momentum is determined.

To determine the effect of this translation, we recall first that the angular momentum about the point A is related to that about the mass center C by

$$\mathbf{H}_A = \mathbf{H}_C + \mathbf{c} \times m\mathbf{v}_{C/A} \qquad\qquad [12\text{-}12]$$

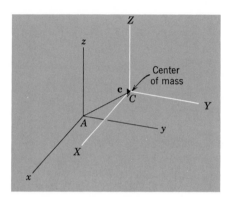

where $\mathbf{c}$ is the position vector locating C relative to A. Noting that $\mathbf{v}_{C/A} = \boldsymbol{\omega} \times \mathbf{c}$ we can write this as

$$\mathbf{H}_A = \mathbf{H}_C + \mathbf{c} \times m(\boldsymbol{\omega} \times \mathbf{c}) = \mathbf{H}_C - m\mathbf{c} \times (\mathbf{c} \times \boldsymbol{\omega})$$

The matrix form of this last equation is

$$[I]_{xyz}\{\omega\} = [I]_{XYZ}\{\omega\} - m[c\times][c\times]\{\omega\}$$

But since this must hold for arbitrary $\{\boldsymbol{\omega}\}$,

$$[I]_{xyz} = [I]_{XYZ} - m[c\times][c\times]$$

Written out in full, this becomes

$$
\begin{bmatrix} I_{xx} & I_{xy} & I_{xz} \\ I_{yx} & I_{yy} & I_{yz} \\ I_{zx} & I_{zy} & I_{zz} \end{bmatrix}
=
\begin{bmatrix} I_{XX} & I_{XY} & I_{XZ} \\ I_{YX} & I_{YY} & I_{YZ} \\ I_{ZX} & I_{ZY} & I_{ZZ} \end{bmatrix}
$$

$$
+ m
\begin{bmatrix} (c_y^2 + c_z^2) & -c_x c_y & -c_x c_z \\ -c_y c_x & (c_z^2 + c_x^2) & -c_y c_z \\ -c_z c_x & -c_z c_y & (c_x^2 + c_y^2) \end{bmatrix}
\qquad (16\text{-}9)
$$

These relationships reveal that each of the products of inertia may be computed by adding, to the corresponding product of inertia with respect to parallel axes through the mass center, a term representing the product of inertia the body would have if the mass were concentrated at C.

Principal Axes of Inertia. As implied by the results in Section 15-4, for a given reference point A, there will be three, mutually perpendicular *principal directions* in any rigid body. Each of these directions has the property that the associated angular momentum vector will coincide with the angular velocity if this vector is in a principal direction. That is, denoting a principal direction by i, we have

$$\mathbf{H}_i = I_i \boldsymbol{\omega}_i$$

or

$$\{H\}_i = I_i \{\omega\}_i$$

Comparing this with Equation 15-11a, we see that the constant I_i is the ith eigenvalue. *Physically*, it is the moment of inertia about the ith principal axis, as can be seen by writing Equation 16-8 for the case in which one of the co-ordinate axes is aligned with $\mathbf{H}_i = I_i \boldsymbol{\omega}_i$.

Example

The products of inertia, with respect to the x,y,z axes of the thin, rectangular plate shown in Figure 16-7a, are

$$\frac{m}{12(a^2 + b^2)} \begin{bmatrix} 2a^2b^2 & -ab(a^2 - b^2) & 0 \\ -ab(a^2 - b^2) & a^4 + b^4 & 0 \\ 0 & 0 & (a^2 + b^2)^2 \end{bmatrix}$$

Figure 16-7

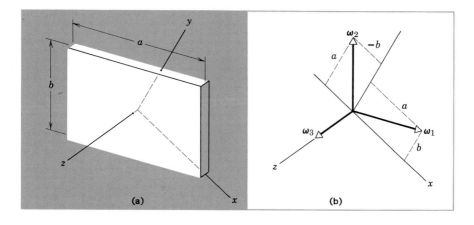

(a)   (b)

Determine the orientation of the principal axes and the moments of inertia about these.

With the eigenvalue denoted by I, the equation (15-12a) becomes, in this case,

$$\begin{vmatrix} \dfrac{ma^2b^2}{6(a^2+b^2)} - I & -\dfrac{mab(a^2-b^2)}{12(a^2+b^2)} & 0 \\[3mm] -\dfrac{mab(a^2-b^2)}{12(a^2+b^2)} & \dfrac{m(a^4+b^4)}{12(a^2+b^2)} - I & 0 \\[3mm] 0 & 0 & \dfrac{m(a^2+b^2)}{12} - I \end{vmatrix} = 0$$

Instead of evaluating the coefficients according to (15-12b), we may take advantage of the presence of the zero elements by expanding by minors of the elements in the third row or column:

$$\left(\frac{m(a^2+b^2)}{12} - I\right) \begin{vmatrix} \dfrac{ma^2b^2}{6(a^2+b^2)} - I & -\dfrac{mab(a^2-b^2)}{12(a^2+b^2)} \\[3mm] -\dfrac{mab(a^2-b^2)}{12(a^2+b^2)} & \dfrac{m(a^4+b^4)}{12(a^2+b^2)} - I \end{vmatrix} = 0$$

Expanding the determinant and simplifying give

$$\left(\frac{m(a^2+b^2)}{12} - I\right)\left(I^2 - \frac{m}{12}(a^2+b^2)I + \left(\frac{mab}{12}\right)^2\right) = 0$$

The two roots from the quadratic factor are

$$I_1, I_2 = \frac{m}{24}(a^2+b^2) \pm \sqrt{\left[\frac{m}{24}(a^2+b^2)\right]^2 - 4\left(\frac{mab}{24}\right)^2}$$

$$= \frac{mb^2}{12}, \quad \frac{ma^2}{12}$$

while the third is

$$I_3 = \frac{m(a^2+b^2)}{12}$$

To find the eigenvector ω_1 corresponding to I_1, we substitute I_1 into the equation equivalent to (15-11):

$$
\begin{bmatrix}
\dfrac{ma^2b^2}{6(a^2+b^2)} - I_1 & -\dfrac{mab(a^2-b^2)}{12(a^2+b^2)} & 0 \\[3ex]
-\dfrac{mab(a^2-b^2)}{12(a^2+b^2)} & \dfrac{m(a^2+b^2)}{12(a^2+b^2)} - I_1 & 0 \\[3ex]
0 & 0 & \dfrac{m(a^2+b^2)}{12} - I_1
\end{bmatrix}
\begin{Bmatrix} \omega_x \\ \omega_y \\ \omega_z \end{Bmatrix}_1
= \begin{Bmatrix} 0 \\ 0 \\ 0 \end{Bmatrix}
$$

With the value $mb^2/12$ inserted for I_1, the three equations implied by the above are

$$
\frac{mb^2(a^2-b^2)}{12(a^2+b^2)}\,\omega_{x1} - \frac{mab(a^2-b^2)}{12(a^2+b^2)}\,\omega_{y1} = 0
$$

$$
-\frac{mab(a^2-b^2)}{12(a^2+b^2)}\,\omega_{x1} + \frac{ma^2(a^2-b^2)}{12(a^2+b^2)}\,\omega_{y1} = 0
$$

$$
\frac{ma^2}{12}\,\omega_{z1} = 0
$$

The last equation gives $\omega_{z1} = 0$, and either of the first two gives $\omega_{x1}/\omega_{y1} = b/a$. Thus, the eigenvector corresponding to I_1 has the components

$$
\{\omega\}_1 = \beta_1 \begin{Bmatrix} a \\ b \\ 0 \end{Bmatrix}
$$

where β_1 is any constant. Similar calculations using the values of I_2 and I_3 give the other two eigenvectors:

$$
\{\omega\}_2 = \beta_2 \begin{Bmatrix} b \\ -a \\ 0 \end{Bmatrix} \qquad
\{\omega\}_3 = \beta_3 \begin{Bmatrix} 0 \\ 0 \\ 1 \end{Bmatrix}
$$

Figure 16-7b illustrates the directions specified by these eigenvectors.

The operation of premultiplication by the original matrix $[I]_{xyz}$ of any one of the vectors represented by $\{\omega\}_1$, $\{\omega\}_2$, or $\{\omega\}_3$ will result in merely "stretching" the vector ω_i. The "stretching" factor is the eigenvalue corresponding to the particular eigenvector on which the operation takes place.

If a coordinate system with axes coinciding with the principal directions had been used instead of those shown, the eigenvectors would have the representations

$$\omega_1: \begin{Bmatrix} \omega_1 \\ 0 \\ 0 \end{Bmatrix} \qquad \omega_2: \begin{Bmatrix} 0 \\ \omega_2 \\ 0 \end{Bmatrix} \qquad \omega_3: \begin{Bmatrix} 0 \\ 0 \\ \omega_3 \end{Bmatrix}$$

The operator that would perform the appropriate "stretching" of these vectors would be represented in this coordinate system by the matrix

$$\begin{bmatrix} I_1 & 0 & 0 \\ 0 & I_2 & 0 \\ 0 & 0 & I_3 \end{bmatrix}$$

You may find it instructive to identify the direction cosine matrix that will transform the x,y,z axes to a set of axes aligned with the principal directions just determined, and apply Equation 15-10 to the given $[I]_{xyz}$.

Kinetic Energy of Rotation.   In Chapter 13, we indicated that the kinetic energy of a moving rigid body can be expressed as

$$T = \tfrac{1}{2} m v_A^2 + m \mathbf{v}_A \cdot \mathbf{v}_{C/A} + \tfrac{1}{2} \int_m v_{P/A}^2 \, dm \qquad [13\text{-}9a]$$

With the velocity differences expressed in terms of the angular velocity,

$$\mathbf{v}_{C/A} = \omega \times \mathbf{r}_{C/A} \qquad \mathbf{v}_{P/A} = \omega \times \mathbf{r}$$

this can also be expressed as

$$T = \tfrac{1}{2} m v_A^2 + m \mathbf{v}_A \cdot (\omega \times \mathbf{r}_{C/A}) + \tfrac{1}{2} \int_m (\omega \times \mathbf{r}) \cdot (\omega \times \mathbf{r}) \, dm$$

Our primary concern here will be with the last term,

$$T_\omega = \tfrac{1}{2} \int_m (\omega \times \mathbf{r}) \cdot (\omega \times \mathbf{r}) \, dm$$

With the help of the vector identity (3-14), this may be written as

$$T_\omega = \tfrac{1}{2}\,\boldsymbol{\omega}\cdot\int_m \mathbf{r}\times(\boldsymbol{\omega}\times\mathbf{r})\,dm$$

or,

$$T_\omega = \tfrac{1}{2}\,\boldsymbol{\omega}\cdot\mathbf{H} \tag{16-10a}$$

$$= \tfrac{1}{2}\,\{\omega\}^T[I]\{\omega\} \tag{16-10b}$$

If the coordinate axes are aligned with the principal axes of inertia, this expression becomes

$$T_\omega = \tfrac{1}{2}(I_1\omega_1{}^2 + I_2\omega_2{}^2 + I_3\omega_3{}^2) \tag{16-10c}$$

Example

The rectangular plate of the previous example is spinning about the x axis at the rate ω. Evaluate its kinetic energy.

The components of angular velocity, referred to the principal directions, are

$$\omega_1 = \frac{a}{\sqrt{a^2+b^2}}\,\omega \qquad \omega_2 = \frac{-b}{\sqrt{a^2+b^2}} \qquad \omega_3 = 0$$

With these values and those of the moments of inertia determined earlier, Equation 16-10 gives

$$T_\omega = \frac{1}{2}\left[\frac{mb^2}{12}\left(\frac{a\omega}{\sqrt{a^2+b^2}}\right)^2 + \frac{ma^2}{12}\left(\frac{-b\omega}{\sqrt{a^2+b^2}}\right)^2\right]$$

$$= \frac{1}{2}\,\frac{ma^2b^2}{6(a^2+b^2)}\,\omega^2$$

It is instructive to verify that the same result may be obtained from Equation 13-9b, using the x,y,z coordinates.

Cauchy's Inertia Ellipsoid. Equations 16-8, 16-9, and 15-10 constitute the necessary computational tools for evaluating angular momentum and kinetic energy of rotation, in various coordinate systems. However, the amount of detail contained in these relationships makes it quite difficult to achieve an overall picture of the inertial properties of a given rigid body. The ingeneous geometric interpretation described in the following fulfills this need beautifully.

Let us plot, in the $\omega_x,\ \omega_y,\ \omega_z$ space, the surface $T_\omega = K$ (constant). That is, allowing the direction (in the body) of angular velocity to vary, we adjust its magnitude such that for each direction we have the same kinetic energy; in this way, the head of the angular velocity vector generates a surface that we shall now examine. The equation for the surface,

$$T_\omega(\omega_x, \omega_y, \omega_z) = \tfrac{1}{2}\{\omega\}^T[I]\{\omega\} = K$$

takes its simplest form when the coordinate axes are the principal axes of inertia:

$$I_1\omega_1{}^2 + I_2\omega_2{}^2 + I_3\omega_3{}^2 = 2K$$

The surface represented by this equation is an ellipsoid with semi-axes

$$a_1 = \sqrt{\frac{2K}{I_1}}$$

$$a_2 = \sqrt{\frac{2K}{I_2}}$$

$$a_3 = \sqrt{\frac{2K}{I_3}}$$

The *size* of the ellipsoid is determined by the value chosen for the constant K, and is of no interest for our present purpose. Its *shape*, however, depends solely on the inertial properties of the body, as we can see from the expression for the ratio between any two semiaxes of the ellipsoid:

$$\frac{a_i}{a_j} = \sqrt{\frac{I_j}{I_i}}$$

A closer examination of this shape lends considerable insight into the inertial properties of the rigid body. Two relationships, developed in the following, provide the keys to a direct visual interpretation of the essential features of the relatively complicated relationship (16-8)

1. The component of angular momentum in the direction of the angular velocity is given by

$$H_\omega = \mathbf{H} \cdot \mathbf{u}_\omega = \mathbf{H} \cdot \frac{\boldsymbol{\omega}}{\omega} = \frac{2K}{\omega}$$

But by placing a coordinate axis in the direction of $\boldsymbol{\omega}$ we can see, from Equation 16-8, that this component of angular momentum may also be written as

$$H_\omega = I_{\omega\omega}\,\omega$$

where $I_{\omega\omega}$ is the moment of inertia about the axis through the reference point A in the direction of $\boldsymbol{\omega}$. Combining the above two expressions for H_ω gives us*

$$I_{\omega\omega} = \frac{2K}{\omega^2} \tag{16-11}$$

Thus, *the moment of inertia about any axis passing through the reference point A is inversely proportional to the square of the distance from the reference point to the inertia ellipsoid.*

2. Consider the gradient (with respect to the coordinates ω_x, ω_y, ω_z) of the function $T_\omega(\omega_x, \omega_y, \omega_z)$:

$$\boldsymbol{\nabla} T = \frac{\partial T}{\partial \omega_x}\,\mathbf{u}_x + \frac{\partial T}{\partial \omega_y}\,\mathbf{u}_y + \frac{\partial T}{\partial \omega_z}\,\mathbf{u}_z$$

This gradient vector has the x component

$$\frac{\partial T}{\partial \omega_x} = \frac{1}{2}\left([1 \quad 0 \quad 0][I]\{\omega\} + \{\omega\}^T[I]\begin{Bmatrix} 1 \\ 0 \\ 0 \end{Bmatrix} \right)$$

$$= I_{xx}\omega_x + I_{xy}\omega_y + I_{xz}\omega_z$$

$$= H_x$$

Similar calculations for the other two components of $\boldsymbol{\nabla} T_\omega$ reveal that

$$\boldsymbol{\nabla} T_\omega = H_x\mathbf{u}_x + H_y\mathbf{u}_y + H_z\mathbf{u}_z = \mathbf{H} \tag{16-12}$$

---

* Note that this is consistent with (13-9b).

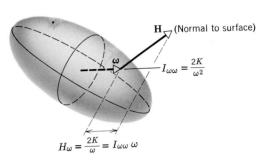

Figure 16-8

Cauchy's inertia ellipsoid.

$$H_\omega = \frac{2K}{\omega} = I_{\omega\omega}\,\omega$$

But since the gradient of a function is normal to the surface defined by a constant value of the function, it follows that *the angular momentum vector is directed perpendicular to the ellipsoid at the point where the related angular velocity meets the surface.* This last result is depicted in Figure 16-8.

Observe that, if two of the moments of inertia about principal axes happen to be equal, the surface is an ellipsoid of revolution and any axis perpendicular to the axis of revolution is a principal axis. The body may possess such inertial symmetry, even though it lacks any corresponding rotational geometric symmetry; an example would be a rectangular parallelepiped with two equal sides, with the reference point A at the center.

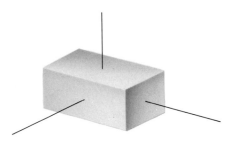

If the moments of inertia about three different axes (not all lying in a plane) are equal, the body is *inertially spherical*, and every axis passing through the reference point is a principal axis.

We will see later how the above properties of the ellipsoid may be used to visualize the rather complicated motion of a rigid body in the absence of torque.

Problems

16-21   Evaluate the x-y-z components of the inertia tensor for the homogeneous rectangular parallelepiped, and determine the principal directions and corresponding moments of inertia.

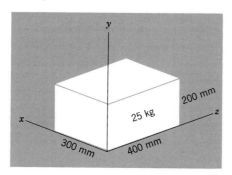

16-22 The homogeneous wedge has a total mass m. Evaluate I_{xy} and I_{xz}.

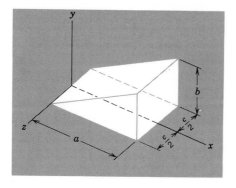

16-23 The slender rod of length $2l$ is oriented with its axis at an angle θ with the x axis.

(a) Determine all components of the inertia tensor $[I]$ by integration (i.e., the basic definition of moment or product of inertia).

(b) Check your answer to (a) by transforming the principal moments of inertia of the thin rod through an angle θ about the proper axis indicated.

(c) Sketch the inertia ellipsoid.

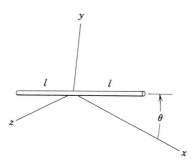

16-24   (a) Sketch the inertia ellipsoid.

(b) The products of inertia with respect to the x-y-z axes are

$$[I] = \begin{bmatrix} 601 & -180 & 0 \\ -180 & 244 & 0 \\ 0 & 0 & 845 \end{bmatrix} \text{mg} \cdot \text{m}^2$$

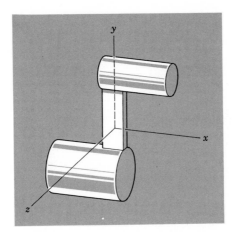

Determine the principal directions and corresponding moments of inertia.

(c) Show the approximate orientation of the angular momentum vector when the body is spinning about the x axis.

16-25   The products of inertia of a certain rigid body, with respect to co-ordinates x,y,z have been found to be

$$[I] = \begin{bmatrix} 9 & 0 & 4 \\ 0 & 7 & 0 \\ 4 & 0 & 3 \end{bmatrix} \text{lb} \cdot \text{in} \cdot \text{s}^2$$

(a) Find the principal directions and corresponding moments of inertia.

(b) Write the coordinate transformation matrix connecting the original axes to the axes aligned with the principal directions.

(c) Use this coordinate transformation in Equation 15-10 to determine the matrix of products of inertia with respect to the principal axes.

16-26   With respect to a set of axes x,y,z a rigid body has products of inertia:

$$[I] = \begin{bmatrix} 10 & 4 & 0 \\ 4 & 4 & 0 \\ 0 & 0 & 1 \end{bmatrix} \text{kg} \cdot \text{m}^2$$

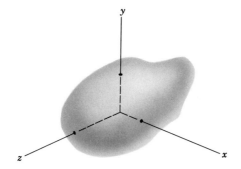

(a) Determine the principal moments of inertia and orientations of the corresponding principal axes.
(b) Sketch the two views of the inertia ellipsoid.

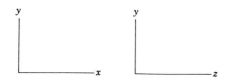

16-27   What shifts of the origin of coordinates will leave the principal directions of inertia unaltered?

16-28   Sketch the inertia ellipsoid for a homogeneous cube,
(a) with the reference point at the mass center, and
(b) with the reference point at a corner.

16-29   (a) Evaluate the products of inertia of the homogeneous cube with respect to the axes shown. Use whatever tabulated values are avilable.
(b) What will be the product of inertia with respect to an axis AB and any axis through A perpendicular to AB?
(c) Determine the principal directions and corresponding moments of inertia.

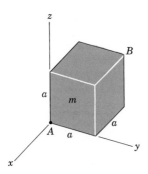

16-30   What is the greatest angle which can be subtended by **H** and **ω**? Suggestion: Consider the equation that gives T_ω in terms of these two vectors.

16-31   Show that the moment of inertia of the three-bladed propeller about x is equal to that about y.

16-32   Show from the integral definitions of moments of inertia that the quantity J_1 (Equation 15-13) is an "invariant" of the inertia tensor; that is, although I_{xx}, I_{yy}, and I_{zz} each depend on the coordinate system, the sum $I_{xx} + I_{yy} + I_{zz}$ does not: $I_{xx} + I_{yy} + I_{zz} = \bar{I}_{xx} + \bar{I}_{yy} + \bar{I}_{zz}$.

16-3

APPLICATION OF M = Ḣ   The basic relationships connecting the forces acting on a rigid body and its motion are

$$\mathbf{f} = m\mathbf{a}_C \qquad\qquad [12\text{-}2]$$

$$\mathbf{M}_A = \dot{\mathbf{H}}_A + m\mathbf{r}_{C/A} \times \mathbf{a}_A \qquad\qquad [12\text{-}13]$$

The application of Equation 12-2 has been illustrated in previous chapters. Also, the proper writing of the term $m\mathbf{r}_{C/A} \times \mathbf{a}_A$ should pose no difficulty by now. We therefore concentrate here on the term $\dot{\mathbf{H}}$.

In evaluating $\dot{\mathbf{H}}$, we must remember that this vector depends *in two ways* on an inertial reference frame. First, the angular momentum vector is defined in terms of the inertial-observed angular velocity of the body. Second, the derivative indicated by the dot is that observed from an inertial reference frame.

Figure 16-9

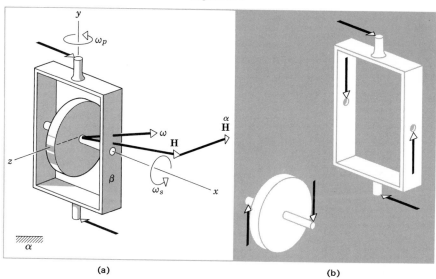

(a)   (b)

For illustration, consider the angular momentum about the center of the flywheel depicted in Figure 16-9a. It is spinning in the gimbal frame with constant angular velocity $\boldsymbol{\omega}_s$, while the gimbal frame rotates around the vertical with angular velocity $\boldsymbol{\omega}_p$ of constant magnitude. Then, according to Equation 14-9, the angular velocity of the flywheel is

$$\boldsymbol{\omega} = \boldsymbol{\omega}_s + \boldsymbol{\omega}_p$$

Because of the symmetry of the flywheel, the directions of these two components of angular velocity are principal directions of inertia, so that the angular momentum is simply

$$\mathbf{H} = I_s\boldsymbol{\omega}_s + I_p\boldsymbol{\omega}_p$$

From the figure it is evident that the component $I_s\boldsymbol{\omega}_s$ will remain fixed in the gimbal frame, but execute a rotation around the vertical relative to an inertial reference frame. Therefore, $\overset{\alpha}{\dot{\mathbf{H}}}$ is in the direction indicated in the figure.

A formal procedure for evaluating $\overset{\alpha}{\dot{\mathbf{H}}}$ in this situation results from considering a reference frame β attached to the gimbal. The angular velocity of this reference frame relative to an inertial reference frame α is then

$$_{\alpha}\boldsymbol{\Omega}_{\beta} = \boldsymbol{\omega}_p$$

Application of Equation 14-1 to the vector $\mathbf{H}$ gives

$$\overset{\alpha}{\mathbf{H}} = \overset{\beta}{\mathbf{H}} + {}_\alpha\boldsymbol{\Omega}_\beta \times \mathbf{H}$$
$$= \mathbf{0} + \boldsymbol{\omega}_p \times (I_s\boldsymbol{\omega}_s + I_p\boldsymbol{\omega}_p)$$
$$= -I_s\omega_s\omega_p\mathbf{u}_z$$

Now, according to Equation 12-13, the moment of forces acting on the flywheel is equal to this last vector. Therefore the gimbal must exert the reactions indicated in the free-body diagrams of Figure 16-9b.

The moment of forces, acting in the direction perpendicular to the two angular velocity components, is somewhat surprising. It is commonly called a *gyroscopic moment*.

More complicated situations are handled most easily with matrix notation. The basic idea illustrated above proceeds as follows.

First, a set of coordinate directions is selected and the angular momentum evaluated according to Equation 16-8.

$$\{H\} = [I]\{\omega\}$$

Now, if these axes are defined to be fixed in an inertial reference frame α, we can write

$$\dot{\mathbf{H}}: \quad {}_\alpha\{\dot{H}\} = [I]\,{}_\alpha\{\dot{\omega}\} + {}_\alpha[\dot{I}]\{\omega\} \tag{16-13a}$$

However, the difficulty with this is that the components of the inertia tensor are varying, because of the rotation of the body relative to the fixed axes. Although these derivatives can be computed* it is normally more straightforward to define the axes to be fixed in a reference frame β, rotating such that the products of inertia do not vary. Then we must use Equation 15-7 to determine the α-observed derivative of $\mathbf{H}$:

$$\dot{\mathbf{H}}: \quad {}_\alpha\{\dot{H}\} = {}_\beta\{\dot{H}\} + [\Omega\times]\{H\}$$
$$= [I]_\beta\{\dot{\omega}\} + [\Omega\times][I]\{\omega\} \tag{16-13b}$$

where $\boldsymbol{\Omega}$ is the angular velocity of the β reference frame relative to the inertial reference frame α.

Example

The antenna depicted in Figure 16-10 is being repositioned. In terms of the angles ϕ and θ and their rates of change, determine the components of moment that must be exerted on the antenna. The moments of inertia about the principal axes x_1, x_2, x_3 are I_1, I_2, and $I_3 = I_2$.

---

* The formula for the rate of change of a second-order tensor, analogous to Equation 15-7, is ${}_\alpha[\dot{I}] = {}_\beta[\dot{I}] + [\Omega\times][I] + [[\Omega\times][I]]^T$. See Problem 15-33.

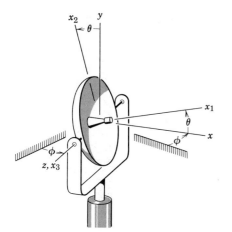

Figure 16-10

Let the antenna be the reference frame β, to which the axes are fixed. Then the angular velocity of the body and of the moving reference frame are equal.

$$_{\alpha}\boldsymbol{\Omega}_{\beta} = \boldsymbol{\omega}: \quad \begin{Bmatrix} \dot{\phi} \sin \theta \\ \dot{\phi} \cos \theta \\ \dot{\theta} \end{Bmatrix}$$

The components of $\overset{\alpha}{\mathbf{H}}$ can now be written, from (16-13b), as

$$_{\alpha}\{\dot{H}\} = \begin{bmatrix} I_1 & 0 & 0 \\ 0 & I_2 & 0 \\ 0 & 0 & I_2 \end{bmatrix} \begin{Bmatrix} \dfrac{d}{dt}(\dot{\phi} \sin \theta) \\ \dfrac{d}{dt}(\dot{\phi} \cos \theta) \\ \ddot{\theta} \end{Bmatrix}$$

$$+ \begin{bmatrix} 0 & -\dot{\theta} & \dot{\phi} \cos \theta \\ \dot{\theta} & 0 & -\dot{\phi} \sin \theta \\ -\dot{\phi} \cos \theta & \dot{\phi} \sin \theta & 0 \end{bmatrix} \begin{bmatrix} I_1 & 0 & 0 \\ 0 & I_2 & 0 \\ 0 & 0 & I_2 \end{bmatrix} \begin{Bmatrix} \dot{\phi} \sin \theta \\ \dot{\phi} \cos \theta \\ \dot{\theta} \end{Bmatrix}$$

After the indicated differentiation and multiplication is carried out, this becomes

$$\overset{\alpha}{\mathbf{H}}: \begin{Bmatrix} I_1\ddot{\phi}\sin\theta + I_1\dot{\phi}\dot{\theta}\cos\theta \\ I_2\ddot{\phi}\cos\theta + (I_1 - 2I_2)\dot{\phi}\dot{\theta}\sin\theta \\ I_2\ddot{\theta} - (I_1 - I_2)\dot{\phi}^2\sin\theta\cos\theta \end{Bmatrix}$$

According to Equation 12-13, the moment of forces acting on the antenna has the x_1, x_2, x_3 components equal to those in this last result. Probably of more interest would be the x,y,z components of moment, because the y and z components would be those required of the driving motors. These could be determined by applying the coordinate transformation

$$\begin{Bmatrix} M_x \\ M_y \\ M_z \end{Bmatrix} = \begin{bmatrix} \cos\theta & -\sin\theta & 0 \\ \sin\theta & \cos\theta & 0 \\ 0 & 0 & 1 \end{bmatrix} \begin{Bmatrix} M_1 \\ M_2 \\ M_3 \end{Bmatrix}$$

$$= \begin{Bmatrix} (I_1 - I_2)\ddot{\phi}\cos\theta\sin\theta + [I_1\cos^2\theta + (2I_2 - I_1)\sin^2\theta]\dot{\phi}\dot{\theta} \\ (I_1\sin^2\theta + I_2\cos^2\theta)\ddot{\phi} + 2(I_1 - I_2)\cos\theta\sin\theta\dot{\phi}\dot{\theta} \\ I_2\ddot{\theta} - (I_1 - I_2)\dot{\phi}^2\sin\theta\cos\theta \end{Bmatrix}$$

If the motion of the antenna is given, then computation of the moment from the above results is straightforward. If, as would more likely be the case, determination of the azimuth and elevation angles $\phi(t)$ and $\theta(t)$ is required in terms of given moment components, we have so far only derived the differential equations of motion. Determination of these angles would require integration of the coupled, nonlinear differential equations

$$(I_1\sin^2\theta + I_2\cos^2\theta)\ddot{\phi} + 2(I_1 - I_2)\cos\theta\sin\theta\,\dot{\phi}\dot{\theta} = M_y(t)$$

$$I_2\ddot{\theta} - (I_1 - I_2)\cos\theta\sin\theta\,\dot{\phi}^2 = M_z(t)$$

Euler's equations express the moment-angular momentum relationship in terms of moment and angular velocity components along the principal axes of inertia. With coordinate axes fixed in the rigid body, as in the above example, the angular velocity of the reference frame is equal to that of the body, so that $\boldsymbol{\Omega}$ may be replaced with $\boldsymbol{\omega}$ in Equation 16-13b. If the reference point A is chosen such that the second term in the right hand side of Equation 12-13 is zero, and if the axes are aligned with the principal axes of inertia, the moment-angular momentum relationship takes the form

$$\begin{Bmatrix} M_1 \\ M_2 \\ M_3 \end{Bmatrix} = \begin{bmatrix} I_1 & 0 & 0 \\ 0 & I_2 & 0 \\ 0 & 0 & I_3 \end{bmatrix} \begin{Bmatrix} \dot{\omega}_1 \\ \dot{\omega}_2 \\ \dot{\omega}_3 \end{Bmatrix}$$

$$+ \begin{bmatrix} 0 & -\omega_3 & \omega_2 \\ \omega_3 & 0 & -\omega_1 \\ -\omega_2 & \omega_1 & 0 \end{bmatrix} \begin{bmatrix} I_1 & 0 & 0 \\ 0 & I_2 & 0 \\ 0 & 0 & I_3 \end{bmatrix} \begin{Bmatrix} \omega_1 \\ \omega_2 \\ \omega_3 \end{Bmatrix}$$

or,

$$M_1 = I_1\dot{\omega}_1 - (I_2 - I_3)\omega_2\omega_3$$
$$M_2 = I_2\dot{\omega}_2 - (I_3 - I_1)\omega_3\omega_1 \qquad (16\text{-}14)$$
$$M_3 = I_3\dot{\omega}_3 - (I_1 - I_2)\omega_1\omega_2$$

These are the famous Euler equations for rigid body motion. You may find it instructive to begin with these to work the above example problems.

Problems

16-33   The shaft is welded to the disc, the axis passing through the center of mass. The unit is spinning in rigid bearings with constant angular velocity.
(a) Evaluate the angular momentum vector and show it on a sketch.
(b) Evaluate the reactions at the bearings and show their directions on a free-body diagram.
(c) Evaluate the kinetic energy of the disc.

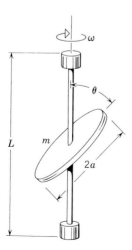

16-34 For the two-throw crankshaft shown, the inertia tensor has been calculated to be

$$I = \begin{bmatrix} 200 & 0 & 40 \\ 0 & 300 & 0 \\ 40 & 0 & 120 \end{bmatrix} \text{lbm} \cdot \text{in}^2$$

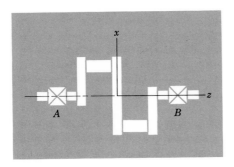

Estimate the magnitude of the bearing reactions due to dynamic unbalance. Indicate both magnitude and direction on a free-body diagram.

16-35 The sketch depicts a type of grinding mill, consisting of a fixed sole, a vertical driving shaft, and conical rollers. The driving shaft turns at angular speed ω_p, and the rollers spin in horizontal bearings so that there is no slippage between the rollers and the sole.
(a) Show on the figure the resultant angular velocity of the right-hand cone.
(b) Evaluate the angular momentum about O, and show it on the figure.
(c) For constant ω_p, evaluate $\dot{\mathbf{H}}_O$.
(d) Assuming that contact is made at the corner at the base of the cone, evaluate the reaction there, indicating it on a free-body diagram.

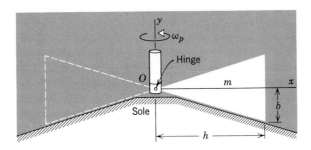

16-36   A crushing device similar to that of Problem 16-35 is shown in the diagram. The lower shaft is hinged to the drive shaft at O, while the disc is free to turn in a bearing at A. The disc rolls around the conical wall without slip.

(a) Show the resultant angular velocity of the disc on the sketch, and determine its magnitude in terms of ω_p and the dimensions given on the diagram.

(b) Evaluate the angular momentum of the disc about O.

(c) For constant ω_p determine the reaction at the wall.

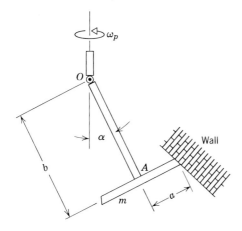

16-37   The disk B has mass m_B, radius R, and spins at ω rad/s relative to the horizontal shaft in the direction shown. The block A has mass m_A and can be moved back and forth on the bar for balancing purposes. Estimate the distance x at which the block should be set for the assembly to precess in the horizontal plane about the Z axis at Ω rad/s.

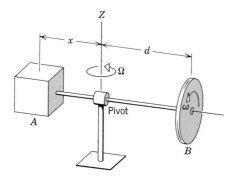

16-38 The essential features of an ideal gyrocompass are depicted in the diagram. A single gimbal is free to rotate about the vertical while constraining the spin axis of the rotor to remain in the horizontal plane. Denote the moments of inertia of the rotor about its spin axis by I_s and its other principal moment of inertia by I_p. Neglect the inertia of the gimbal.

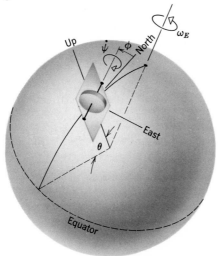

In terms of these quantities and those shown on the diagram, write the differential equations of motion governing ϕ and ψ. Under what condition can we say that $\dot{\psi}$ will be constant? With $\dot{\psi}$ constant, what will be the period of small oscillations in ϕ?

16-39 The T-shaped bar is one solid piece with negligible mass, mounted in bearings that are aligned horizontally. The disc mounted on the lower

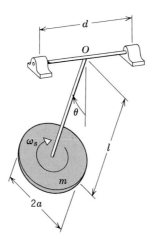

end spins with a constant angular velocity ω_s relative to the shaft. The device oscillates as a pendulum, as indicated by $\theta(t) = \theta_0 \sin pt$.

(a) Evaluate and show on a sketch the angular momentum of the disc about O.

(b) Evaluate and show on a sketch $\dot{\mathbf{H}}_O$.

(c) Evaluate and show on a free-body diagram the reactions from the bearings.

16-40 The compressor-turbine spool inside the turbojet aircraft is spinning at 12,000 rpm in the direction indicated, as the aircraft performs a tight turn. On a sketch of the spool as a free-body diagram, indicate the directions of the reactions at the bearings at A and B, induced by the rotor spin and aircraft maneuver. (Neglect the reaction due to gravity.)

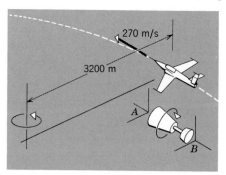

16-41 Estimate the moment (magnitude and direction) that the rotor shaft will exert on the helicopter as a result of the helicopter rotating about a horizontal axis as shown at an approximately constant rate of 90° per second. The mass of the rotor is 230 kg, its radius of gyration is 6.1 m, and it spins at 100 rpm (in a clockwise direction viewed from above).

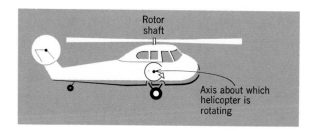

16-42 Special washers are used to tilt the saw blade as shown in order to cut a wide groove in the work. The moments of inertia of the blade are

$$I_{yy} = 0.015 \text{ lb·in·s}^2$$

$$I_{zz} = 0.030 \text{ lb·in·s}^2$$

At 3600 rpm, what moment will be exerted on the shaft when the blade is not in contact with the work?

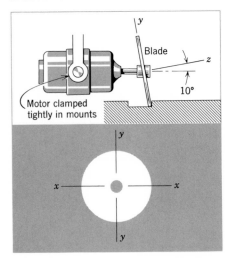

16-4

THE WORK-KINETIC ENERGY INTEGRAL FOR A RIGID BODY.
We have already deduced general expressions for the work of the forces acting on a rigid body (Equation 13-8), and for the kinetic energy (Equations 13-9b and 16-10. Since a rigid body may be considered as a special system of particles, in which the distance between every pair of particles remains fixed, the work-kinetic energy relationship follows directly from Equation 12-15b. It is instructive to follow an alternate course, beginning with the force-acceleration laws for the rigid body.

Dot multiplication of each member of Equation 12-2 with the increment of displacement of the reference point A yields

$$\mathbf{f} \cdot d\mathbf{r}_A = m\mathbf{v}_C \cdot \mathbf{v}_A \, dt$$

$$= m \frac{d}{dt} (\mathbf{v}_A + \mathbf{v}_{C/A}) \cdot \mathbf{v}_A \, dt$$

$$= m\mathbf{v}_A \cdot d\mathbf{v}_A + m\mathbf{v}_A \cdot d\mathbf{v}_{C/A} \tag{a}$$

Dot multiplication of each member of Equation 12-13 with the increment of angular displacement yields

$$\mathbf{M}_A \cdot d\mathbf{\Phi} = (\dot{\mathbf{H}}_A + m\mathbf{r}_{C/A} \times \dot{\mathbf{v}}_A) \cdot \mathbf{\omega} \, dt$$

$$= \dot{\mathbf{H}}_A \cdot \mathbf{\omega} \, dt + m(\mathbf{\omega} \times \mathbf{r}_{C/A}) \cdot \dot{\mathbf{v}}_A \, dt$$

$$= d\mathbf{H}_A \cdot \mathbf{\omega} + m\mathbf{v}_{C/A} \cdot d\mathbf{v}_A \tag{b}$$

In order to relate the first term on the right-hand side of (b) to the kinetic energy of rotation, we need to employ the following relationship:

$$\boldsymbol{\omega} \cdot \dot{\mathbf{H}} = \dot{\boldsymbol{\omega}} \cdot \mathbf{H} \tag{c}$$

To demonstrate the validity of this equation, consider first the rate of change, as observed from a reference frame fixed to the body itself, of the angular momentum. With coordinate axes fixed to the body, this vector will have the resolution

$$\overset{\beta}{\mathbf{H}}: {}_{\beta}\{\dot{H}\} = [I] {}_{\beta}\{\dot{\omega}\}$$

and the dot product of this vector with the angular velocity has the value

$$\boldsymbol{\omega} \cdot \overset{\beta}{\mathbf{H}} = \{\omega\}^{T}{}_{\beta}\{\dot{H}\} = \{\omega\}^{T} [I]{}_{\beta}\{\dot{\omega}\}$$

Now, it is easy to verify by writing out the individual terms in the sum, that because of the symmetry of the matrix $[I]$, interchange of the row and column will not affect the value of the matrix product:

$$\{\omega\}^{T} [I]{}_{\beta}\{\dot{\omega}\} = {}_{\beta}\{\dot{\omega}\}^{T} [I]\{\omega\}$$

That is,

$$\boldsymbol{\omega} \cdot \overset{\beta}{\mathbf{H}} = \overset{\beta}{\dot{\boldsymbol{\omega}}} \cdot \mathbf{H}$$

But the body-observed rates of change are related to the inertial-observed rates of change through Equation 14-1; thus we may write

$$\boldsymbol{\omega} \cdot (\overset{\alpha}{\dot{\mathbf{H}}} - \boldsymbol{\omega} \times \mathbf{H}) = (\overset{\alpha}{\dot{\boldsymbol{\omega}}} - \boldsymbol{\omega} \times \boldsymbol{\omega}) \cdot \mathbf{H}$$

which reduces to the desired relationship (c).

Applying this, and the fact that the derivative of the scalar $\mathbf{A} \cdot \mathbf{B}$ is independent of reference frame, we can show that the first term on the right-hand side is equal to the increment of increase in rotational kinetic energy:

$$(d\mathbf{H}) \cdot \boldsymbol{\omega} = \frac{(d\mathbf{H}) \cdot \boldsymbol{\omega} + (d\boldsymbol{\omega}) \cdot \mathbf{H}}{2} = d(\tfrac{1}{2}\boldsymbol{\omega} \cdot \mathbf{H}) \tag{d}$$

With this result, addition of Equations a and b gives the work-kinetic energy relationship for a rigid body:

$$\mathbf{f} \cdot d\mathbf{r}_A + \mathbf{M}_A \cdot d\boldsymbol{\Phi} = m\mathbf{v}_A \cdot d\mathbf{v}_A$$
$$+ m(\mathbf{v}_A \cdot d\mathbf{v}_{C/A} + \mathbf{v}_{C/A} \cdot d\mathbf{v}_A) + d(\tfrac{1}{2}\boldsymbol{\omega} \cdot \mathbf{H}_A)$$
$$= d\left(\frac{m\mathbf{v}_A{}^2}{2} + m\mathbf{v}_A \cdot \mathbf{v}_{C/A} + \frac{\boldsymbol{\omega} \cdot \mathbf{H}_A}{2} \right) \tag{16-15a}$$

Or,

$$dW = dT$$

A review of the above derivation, with the reference point chosen at the mass center, will show that for this case the separate relationships

$$\mathbf{f} \cdot d\mathbf{r}_C = d\,\frac{mv_C^2}{2} = d(T_{\text{transl.}}) \qquad (16\text{-}15b)$$

$$\mathbf{M}_C \cdot d\mathbf{\Phi} = d\,\frac{\boldsymbol{\omega} \cdot \mathbf{H}_C}{2} = d(T_{\text{rot.}}) \qquad (16\text{-}15c)$$

are valid.

Problems

16-43   Use the fact that T_ω must be independent of coordinate system,

$$\tfrac{1}{2}\{\overline{\omega}\}^T[\overline{I}]\{\overline{\omega}\} = \tfrac{1}{2}\{\omega\}[I]\{\omega\}$$

to deduce the transformation formula

$$[\overline{I}] = [l][I][l]^T$$

16-44   Let the reference point A be on the screw axis. (See pp. 40–41.)
(a) Show that with this choice, the translation-rotation coupling term $m\mathbf{v}_A \cdot (\boldsymbol{\omega} \times \mathbf{r}_{C/A})$ in (13-9a), p. 106, vanishes.
(b) Show from this that for two-dimensional motion the kinetic energy may be expressed as

$$T = \tfrac{1}{2}I_{C'}\omega^2$$

where $I_{C'}$ is the moment of inertia about the instantaneous axis of rotation.
(c) Verify the result from (b) by letting A be the mass center C in (13-9a), and relating the speed of C to its distance from the instantaneous axis and the angular velocity ω.
(d) Is $\mathbf{M}_{C'} \cdot \boldsymbol{\omega}\, dt = d(\tfrac{1}{2}\mathbf{H}_{C'} \cdot \boldsymbol{\omega})$?

16-5

SOME ANALYTICALLY SOLVED PROBLEMS. Problems of predicting the motions of systems with rigid bodies typically lead to sets of differential equations that cannot be integrated analytically. However, some insight can be gained by examining some of the simpler situations in which analytical integration and, consequently, a thorough understanding of the dynamic behavior, have been achieved.

Torque-Free Motion.   The motion of an unsymmetric rigid body, in the absence of torque about the center of mass, is more complex than might be at first estimated. Any good dynamics text will reveal some of the intricacies of such motion; fasten the book closed with a stiff rubber band and observe it closely after tossing it in the air spinning as indicated in the sketch. Classical studies of torque-free motion have produced a solution giving the orientation of the body as a function of time, and an ingeneous geometrical mechanism for describing the motion. The analytical solution is presented in Whittaker's book.* The mechanism of L. Poinsot provides a simpler vehicle for gaining an understanding of the motion and so will be presented here.

As shown in Figure 16-11a, a plane is fixed perpendicular to the angular momentum vector, at a distance from the center of mass equal to the projection of ω onto $\mathbf{H}$. The rotation of the body is such that its inertia ellipsoid rolls without slipping on this "invariable plane," with the center of mass C fixed relative to the plane.

To see that such rotation satisfies the laws of motion, note first that because the torque is zero, the angular momentum vector $\mathbf{H}$, and consequently the orientation of the invariable plane, are fixed in an inertial reference frame. In the absence of torque the kinetic energy of rotation is also constant, so that as the angular velocity vector varies its orientation in the body, its magnitude varies in such a way that the head of the arrow remains on the inertia ellipsoid. The fixed distance from C to the invariable plane is then a further consequence of the constant values of $\mathbf{H}$ and T_ω, since

$$T_\omega = \tfrac{1}{2}\omega \cdot \mathbf{H}$$

implies

$$\omega \cos \angle_\omega^{\mathbf{H}} = \frac{2T_\omega}{H}$$

* E. T. Whittaker, *A Treatise on the Analytical Dynamics of Particles and Rigid Bodies*, 4th Ed., Dover Publ., 1944.

Figure 16-11

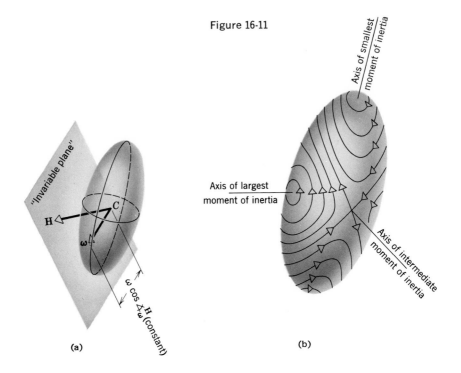

(a)

(b)

Because the angular momentum vector is directed perpendicular to the ellipsoid at the point where the angular velocity vector meets the surface, the ellipsoid and invariable plane are tangent at this point. Finally, because this point lies on the instantaneous axis of rotation, it has zero velocity, so that the ellipsoid rolls without slip.

The path traced out on the ellipsoid by the point of contact is called a *polhode*. Such a curve may be plotted from the equations expressing constant values of kinetic energy and magnitude of angular momentum:

$$I_1\omega_1^2 + I_2\omega_2^2 + I_3\omega_3^2 = 2T$$
$$(I_1\omega_1)^2 + (I_2\omega_2)^2 + (I_3\omega_3)^2 = H^2$$

Or, by carefully considering the mechanism of Poinsot, the form of the curve may be visualized directly. A family of polhodes is illustrated in Figure 16-11*b*. These reveal the most important properties of the torque-free motion.

If the angular velocity exactly coincides with one of the principal axes of inertia, the axis of spin remains fixed in both the body and an inertial reference frame. We can think of these three states of motion as "equilibrium" states; for any other state of motion, the angular velocity will vary periodically in the body. (Its variation in an inertial reference frame is not necessarily periodic.)

A point representing a state of motion is said to be stable if any point that is initially within a neighborhood of the point in question remains for all subsequent time within a neighborhood of the point in question. We can see from Figure 16-11*b* that two of the equilibrium points are stable and the other is unstable. This situation may be easily observed by tossing the rubber-band sealed book.

In Chapter 18 we examine analytically the motions in the neighborhoods of the three equilibrium points.

The torque-free motion of a rigid body having an axis of symmetry is, as would be expected, simpler than that of the body with three different principal moments of inertia. In this case the polhodes reduce to circles with centers on the symmetry axis, as shown in Figure 16-12*a*. Another mechanism, equivalent to

Figure 16-12

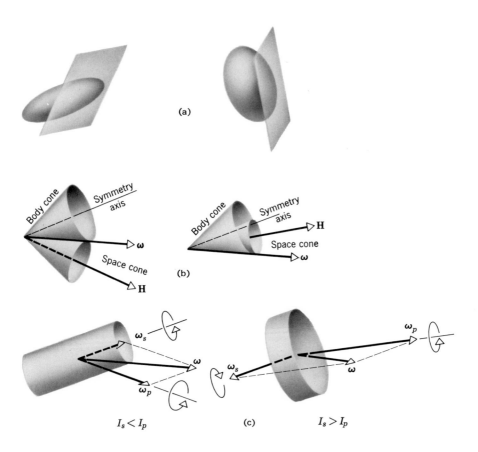

(a)

Body cone   Symmetry axis   Space cone ω **H**

Body cone   Symmetry axis   Space cone ω **H**

(b)

ω_s ω ω_p

ω_s ω ω_p

$I_s < I_p$ (c) $I_s > I_p$

that of Poinsot for the symmetric body, is shown in Figure 16-12*b*. In this mechanism a "space cone" is fixed in an inertial reference frame with the angular momentum vector along its axis, and the angular velocity vector forming its generators. A "body cone" is fixed in the body, its axis coinciding with the axis of inertial symmetry. The motion of the body is such that the body cone rolls without slip around the space cone.

The components of angular velocity along the axes of the body cone and the space cone (i.e., along the symmetry axis of the body and the fixed direction of the angular momentum vector) are called the spin and precession, respectively. These are denoted by ω_s and ω_p and shown in Figure 16-12*c*. When the angle between ω_s and ω_p is acute, the swinging of the symmetry axis around the fixed direction of **H** is termed *direct precession*; when this angle is obtuse the swinging is termed *retrograde precession*.

The Spinning Top. The motion of a symmetrical body with one point fixed, under the influence of a constant gravitational force, presents one of the classical problems of rigid body dynamics. To be determined is the orientation of the symmetry axis, as a function of time and the physical parameters and initial conditions which set the body in motion. The Euler angles ϕ and χ, shown in Figure 16-13, define the orientation of the symmetry axis; the third Euler angle, ψ, is of less interest, but its derivative is the spin component of angular velocity. The components $\dot{\phi}$ and $\dot{\chi}$ are called precession and nutation, respectively.

The differential equations governing these Euler angles will of course express $\mathbf{M} = \dot{\mathbf{H}}$. In order to write these equations, let us resolve the vectors

Figure 16-13

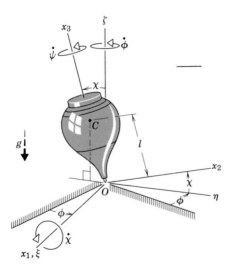

involved onto the x_1, x_2, x_3 axes. The moment about the pivot point has the components

$$\mathbf{M_0}: \begin{Bmatrix} mgl \sin \chi \\ 0 \\ 0 \end{Bmatrix}$$

The products of inertia with respect to the x axes

$$\begin{bmatrix} I_{11} & I_{12} & I_{13} \\ I_{21} & I_{22} & I_{23} \\ I_{31} & I_{32} & I_{33} \end{bmatrix} = \begin{bmatrix} I_1 & 0 & 0 \\ 0 & I_1 & 0 \\ 0 & 0 & I_s \end{bmatrix}$$

will remain constant. The angular velocity of the body, and that of the reference frame β to which the x axes are attached, have the resolutions

$$\boldsymbol{\omega}: \begin{Bmatrix} \dot{\chi} \\ \dot{\phi} \sin \chi \\ \dot{\phi} \cos \chi + \dot{\psi} \end{Bmatrix} \quad \text{and} \quad \boldsymbol{\Omega}: \begin{Bmatrix} \dot{\chi} \\ \dot{\phi} \sin \chi \\ \dot{\phi} \cos \chi \end{Bmatrix}$$

The relationship

$$\mathbf{M_0} = \overset{\beta}{\mathbf{H}_0} + \boldsymbol{\Omega} \times \mathbf{H_0} \qquad \{M\} = [I]_\beta \{\dot{\omega}\} + [\Omega][I]\{\omega\}$$

then takes the form

$$\begin{Bmatrix} mgl \sin \chi \\ 0 \\ 0 \end{Bmatrix} = \begin{Bmatrix} I_1 \dfrac{d}{dt}(\dot{\chi}) \\ I_1 \dfrac{d}{dt}(\dot{\phi} \sin \chi) \\ I_s \dfrac{d}{dt}(\dot{\phi} \cos \chi + \dot{\psi}) \end{Bmatrix}$$

$$+ \begin{bmatrix} 0 & -\dot{\phi} \cos \chi & \dot{\phi} \sin \chi \\ \dot{\phi} \cos \chi & 0 & -\dot{\chi} \\ -\dot{\phi} \sin \chi & \dot{\chi} & 0 \end{bmatrix} \begin{Bmatrix} I_1 \dot{\chi} \\ I_1(\dot{\phi} \sin \chi) \\ I_s(\dot{\phi} \cos \chi + \dot{\psi}) \end{Bmatrix}$$

With the differentiation and multiplication carried out, these become

$$I_1\ddot{\chi} - (I_1 - I_s)\dot{\phi}^2 \sin \chi \cos \chi + I_s\dot{\psi}\dot{\phi} \sin \chi - mgl \sin \chi = 0 \qquad \text{(16-16a)}$$

$$I_1(\ddot{\phi} \sin \chi + 2\dot{\phi}\dot{\chi} \cos \chi) - I_s(\dot{\phi} \cos \chi + \dot{\psi})\dot{\chi} = 0 \qquad \text{(16-16b)}$$

$$I_s \frac{d}{dt} (\dot{\phi} \cos \chi + \dot{\psi}) = 0 \qquad \text{(16-16c)}$$

Integration of these equations in a general way will not be undertaken here. We see in Chapter 17 how an alternate derivation will lead naturally to an integration of the system.

An "equilibrium" motion, consisting of a steady spinning and steady precession about the vertical, can be readily seen to satisfy the laws of motion. In such a motion, the angular momentum vector will have a fixed magnitude and inclination with the vertical, and swing at a constant rate about the vertical; its rate of change will thus remain in the ξ direction, the direction of the gravitational moment. In terms of the above formulation, the motion will be given by constant values of χ, $\dot{\phi}$, and $\dot{\psi}$. Substitution into the above differential equations yields the relationship

$$\left[\dot{\psi}_0 - \left(\frac{I_1}{I_s} - 1\right) \dot{\phi}_0 \cos \chi_0 \right] \dot{\phi}_0 = \frac{mgl}{I_s}$$

and verifies that the constant values satisfy the laws of motion. Unless the body is set in motion in such a way as to satisfy this relationship, we can expect the motion to be more complicated.

Further analysis shows that if the initial conditions *nearly* satisfy the above relationship, the spin axis will execute a small oscillatory motion superimposed on the "equilibrium" motion. These oscillatory variations in $\dot{\phi}$ and χ are called secondary precession and nutation, respectively.

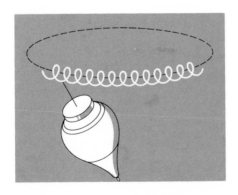

Problems

16-45   A rigid body has an axis of inertial symmetry, about which its moment of inertia is $I_s = 200$ mg·m². Its moment of inertia about a lateral axis through the center of mass is $I_1 = 300$ mg·m². The body is rotating in the absence of moment about its mass center, its spin component of angular velocity equal to 60 rad/s. Its kinetic energy of rotation is $T = 0.360$ J.
(a) Determine the angles of the space and body cones.
(b) Determine the rate at which the symmetry axis will precess.

16-46   Same as Problem 16-45 except that $I_1 = 180$ mg·m².

VIRTUAL WORK FOR DYNAMIC SYSTEMS

The principle of virtual work explained in Section 7-3 is effective for handling problems in dynamics as well as statics. Its use in dynamics requires that inertia forces be included along with the interaction forces as considered in Chapter 7. An orderly procedure for doing this is provided by a remarkable analysis conceived by the famous French mathematician J. L. Lagrange (1736–1813). The objective of this chapter is to develop a working understanding of Lagrange's equations.

17-1

DEGREES OF FREEDOM AND GENERALIZED COORDINATES.
The idealization that objects are rigid imposes limitations on the possible motions of a mechanical system; these are called the *constraints* of the system. For example, we may idealize the guide rod and the pendulum rod in Figure 17-1 as perfectly rigid, with the result that the particle number 1 must remain on the straight line OP, and the particle number 2 must remain at a distance l from particle number 1.

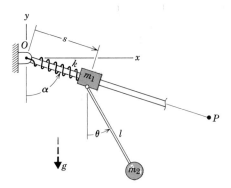

Figure 17-1

If the angle α is given and the motions are limited to the plane of the figure, the system has two *degrees of freedom*, because exactly two quantities, such as s and θ, are required to locate all particles within the system. A system with n degrees of freedom is one in which exactly n quantities are required to uniquely specify the position of every mass particle within the system, for an arbitrary configuration satisfying the constraints. Any set of n such quantities forms a set of *generalized coordinates* for the system.

The constrains in Figure 17-1 might be expressed in the form

$$x_1^2 + y_1^2 = s^2$$
$$(x_1 - x_2)^2 + (y_1 - y_2)^2 = l^2$$

in which (x_1,y_1) and (x_2,y_2) are the rectangular Cartesian coordinates locating particles 1 and 2, respectively. Or, the constraints may be incorporated into the *transformation to generalized coordinates*,

$$\mathbf{r}_1(s,\theta,t) = s \sin \alpha \, \mathbf{u}_x - s \cos \alpha \, \mathbf{u}_y$$
$$\mathbf{r}_2(s,\theta,t) = (s \sin \alpha + l \sin \theta)\mathbf{u}_x - (s \cos \alpha + l \cos \theta)\mathbf{u}_y \qquad \text{(a)}$$

in which $\mathbf{r}_k$ is a position vector locating the kth particle. For a system with p particles and n degrees of freedom, the transformation to generalized coordinates has the form*

$$\mathbf{r}_k = \mathbf{r}_k(q_1, q_2, \cdots, q_n, t) \qquad k = 1, 2, \cdots, p \qquad \text{(17-1)}$$

in which the q_i are the generalized coordinates. Time t does not always appear explicitly as indicated in Equations a and 17-1. If, for example, the angle α in Figure 17-1 were fixed, t would be contained only implicitly in the variables

---

* There are systems having constraints that cannot be expressed in the form (17.1). Those with contraints that can be so expressed are called *holonomic* systems. Approaches to analysis of some nonholonomic systems are discussed in Goldstein, *Classical Mechanics*, Addison-Wesley, 1950 and Kane, "Dynamics of Nonholonomic Systems," ASME *J. Appl. Mechanics*, December, 1961, pp. 574–578.

$s(t)$ and $\theta(t)$. However, if α were some prescribed function of time, this would be reflected in the explicit appearance of t, indicated as the last in the list of variables in (a) and (17-1). Equation 17-1 forms the basis for the development of Lagrange's equations.

Problems

For Problems 17-1 through 17-20, indicate the number of degrees of freedom of each system and which systems have time-dependent constraints (i.e., for which t will appear explicitly in the transformation to generalized coordinates). For Problems 17-1 through 17-6, define two sets of generalized coordinates.

17-1

17-2

17-3

17-4

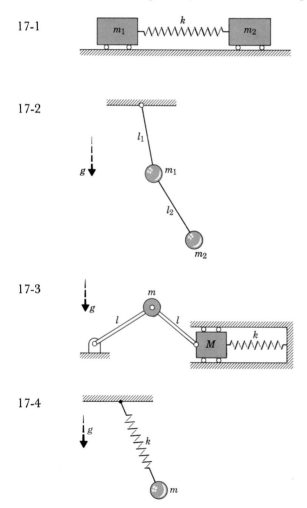

17-5

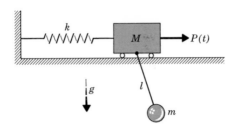

17-6

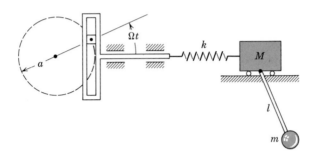

17-7

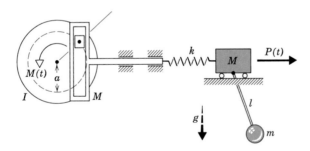

17-8

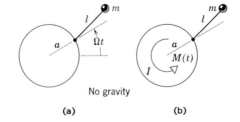

17-9

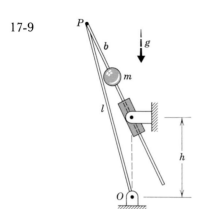

17-10

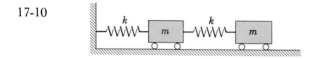

17-11

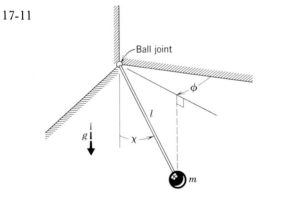

17-12

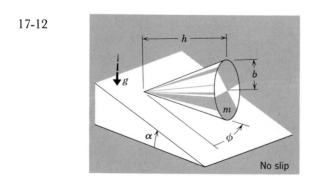

17-13

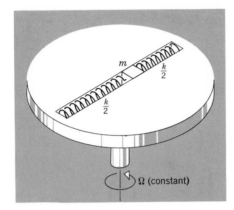

17-14

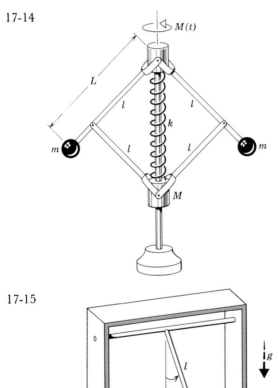

17-15

Ω constant

17-16

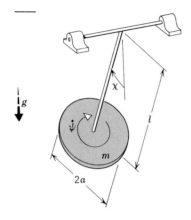

17-17

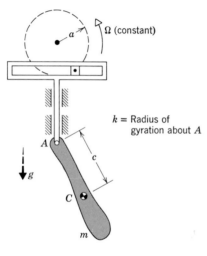

k = Radius of
gyration about A

17-18

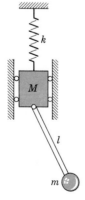

17-19

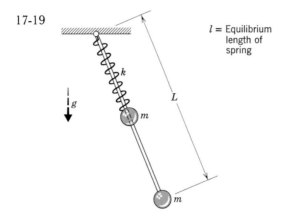

l = Equilibrium
length of
spring

17-20   The gyrocompass of Problem 16-38.

17-2

GENERALIZED FORCE COMPONENTS. In this chapter, we are primarily interested in the expression of Newton's laws of motion in a form in which forces of constraint do not appear. This may be accomplished by considering only virtual displacements that do not violate any constraints, as was done in the first example in Section 7-3. For this reason, we will not consider virtual displacements such as that used in the second example of Section 7-3, in which the length of the rigid connecting rod was imagined to vary.

The analytical treatment of virtual work can be carried out in the manner first explained in Section 7-3; that is, virtual displacements evaluated by formal differentiation of the geometric relationships, which are now expressed by Equation 17-1. For the system of Figure 17-1, differentiation of Equations a results in

$$\delta \mathbf{r}_1 = \frac{\partial \mathbf{r}_1}{\partial s} \, \delta s + \frac{\partial \mathbf{r}_1}{\partial \theta} \, \delta \theta$$

$$= (\sin \alpha \, \delta s) \mathbf{u}_x - (\cos \alpha \, \delta s) \mathbf{u}_y$$

$$\delta \mathbf{r}_2 = \frac{\partial \mathbf{r}_2}{\partial s} \, \delta s + \frac{\partial \mathbf{r}_2}{\partial \theta} \, \delta \theta$$

$$= (\sin \alpha \, \delta s + l \cos \theta \, \delta \theta) \mathbf{u}_x - (\cos \alpha \, \delta s - l \sin \theta \, \delta \theta) \mathbf{u}_y \qquad \text{(b)}$$

If friction is negligible, the only interaction forces that can do work on this virtual displacement are the forces from gravity and the spring,

$$\mathbf{f}_1 = -k(s - s_0) \sin \alpha \, \mathbf{u}_x + [k(s - s_0) \cos \alpha - m_1 g]\mathbf{u}_y$$

$$\mathbf{f}_2 = -m_2 g \mathbf{u}_y$$

The virtual work is then

$$\delta W = \mathbf{f}_1 \cdot \delta \mathbf{r}_1 + \mathbf{f}_2 \cdot \delta \mathbf{r}_2$$

$$= [-k(s - s_0) + (m_1 + m_2)g \cos \alpha]\delta s - m_2 g l \sin \theta \, \delta\theta \qquad \text{(c)}$$

Application of this procedure to the more general case is as follows:

$$\delta \mathbf{r}_k = \sum_{i=1}^{n} \frac{\partial \mathbf{r}_k}{\partial q_i} \delta q_i$$

$$\delta W = \sum_{k=1}^{p} \mathbf{f}_k \cdot \delta \mathbf{r}_k$$

$$= \sum_{k=1}^{p} \mathbf{f}_k \cdot \sum_{i=1}^{n} \frac{\partial \mathbf{r}_k}{\partial q_i} \delta q_i$$

$$= \sum_{i=1}^{n} \sum_{k=1}^{p} \mathbf{f}_k \cdot \frac{\partial \mathbf{r}_k}{\partial q_i} \delta q_i$$

$$\boxed{= \sum_{i=1}^{n} Q_i \delta q_i} \qquad \text{(17-2)}$$

The coefficients

$$Q_i = \sum_{k=1}^{p} \mathbf{f}_k \cdot \frac{\partial \mathbf{r}_k}{\partial q_i} \qquad \text{(17-3)}$$

in Equation 17-2 are called the *generalized force components* corresponding to the coordinates q_i. For the system and coordinates of Figure 17-1, the generalized force components may be taken as the coefficients appearing in Equation c:

$$S = -k(s - s_0) + (m_1 + m_2)g \cos \alpha$$

$$\Theta = -m_2 g l \sin \theta \qquad \text{(d)}$$

As an alternative to the formal differentiation implied in Equation 17-3, it is often easier to evaluate the virtual work and generalized force components by direct examination of the geometry of the small displacements, as explained in Chapter 7.* Observe that the ith term in the expansion (17-2) represents the virtual work that would be done on a virtual displacement consisting of a variation in q_i with all other δqs equal to zero. Thus we can determine Q_i by evaluating the virtual work associated with a small increase in q_i alone. For example, the generalized force components for the system in Figure 17-1 may be determined as follows:

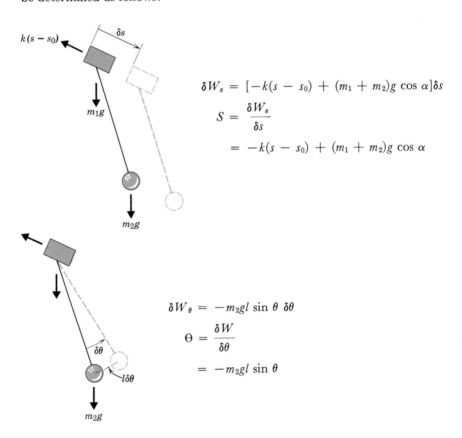

$$\delta W_s = [-k(s - s_0) + (m_1 + m_2)g \cos \alpha]\delta s$$

$$S = \frac{\delta W_s}{\delta s}$$

$$= -k(s - s_0) + (m_1 + m_2)g \cos \alpha$$

$$\delta W_\theta = -m_2 g l \sin \theta \, \delta\theta$$

$$\Theta = \frac{\delta W}{\delta\theta}$$

$$= -m_2 g l \sin \theta$$

Conservative Forces. A generalized force is said to be *conservative* if the components are functions only of the coordinates and t,

$$Q_i = Q_i(q_1, q_2, \cdots, q_n, t) \tag{17-4a}$$

---

* Charles Smith, *Statics*, Wiley, 1976, pp. 180–182.

such that there exists a *potential function* $V(q_1, q_2, \cdots, q_n, \dot{t})$, related to the force by

$$Q_i = -\frac{\partial V}{\partial q_i} \tag{17-4b}$$

The potential function may be determined in terms of the force components by the line integral

$$V = -\int_{q_0}^{q} \sum_i Q_i \, dq_i$$

in which the single letter q in each limit is an abbreviation for the point with coordinates $(q_1, q_2, \cdots, q_n)$. The choice of the lower limit is arbitrary, which means that the potential function may be determined only to within an arbitrary additive constant.

If V is to depend only on the coordinates and t, and not on the path of integration, the integrand $\Sigma Q_i \, dq_i$ must be a perfect differential of V. That is, the force components must satisfy the $n(n-1)/2$ relationships

$$\frac{\partial Q_i}{\partial q_j} = \frac{\partial Q_j}{\partial q_i} \tag{17-4c}$$

throughout a simply connected region containing the possible integration paths.*

In the system of Figure 17-1, this condition,

$$\frac{\partial S}{\partial \theta} = \frac{\partial \Theta}{\partial s}$$

is readily verified from Equations d. The potential function may then be determined by integration along any path, as

$$V = -\int_{(s_0,0)}^{(s,0)} [-k(s - s_0) + (m_1 + m_2)g \cos \alpha] ds$$

$$-\int_{(s,0)}^{(s,\theta)} -m_2 gl \sin \theta \, d\theta$$

$$= \tfrac{1}{2}k(s - s_0)^2 - (m_1 + m_2)g \cos \alpha (s - s_0)$$

$$+ m_2 gl(1 - \cos \theta)$$

* See, for instance, H. Flanders, *Differential Forms*, Academic Press, 1963.

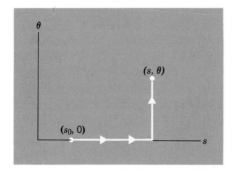

The conservative forces that arise most commonly in engineering applications are from gravity and perfectly elastic deformations. In these cases, it is usually easier to write down the expression for V in terms of known expressions for potential energy, and determine the Q_i by means of Equation 17-4b, rather than by the integration procedure just illustrated.

Note that when the constraints are time dependent (such as the case $\alpha = \alpha(t)$ in the system of Figure 17-1), mechanical energy is not necessarily conserved within the system, even though the force is classified as conservative. When writing the expression for $V(q,t)$ in such cases, we consider potential energy variations with variations in the qs as if the constraint were fixed. This is consistent with the definition of virtual displacements

$$\delta \mathbf{r}_k = \sum_{i=1}^{n} \frac{\partial \mathbf{r}_k}{\partial q_i} \delta q_i$$

in which there is no variation in t. For comparison, an actual dynamic variation in configuration would be expressed by

$$d\mathbf{r}_k = \sum_{i=1}^{n} \frac{\partial \mathbf{r}_k}{\partial q_i} dq_i + \frac{\partial \mathbf{r}_k}{\partial t} dt$$

Friction Forces.   A commonly encountered type of nonconservative force is that associated with friction. For example, suppose a significant tangential component of the reaction between the slider and guide rod in the device shown in Figure 17-1 acts in a direction opposing the sliding motion. With the magnitude of this component denoted by f_t, the virtual work of this force will be

$$\delta W_t = - \frac{\dot{s}}{|\dot{s}|} f_t \, \delta s$$

and a term $-(\dot{s}/|\dot{s}|)f_t$ would be added to the generalized force component S in Equations d.

If the friction force is a function of $\dot{s}$ only,

$$\frac{\dot{s}}{|\dot{s}|} f_t = F(\dot{s})$$

the use of Lagrange's equations will readily incorporate this effect in the equations of motion. However, if we are dealing with Coulomb friction,*

$$f_t = \mu f_n$$

the two Lagrange equations will contain the unknown normal force f_n in addition to the coordinates s and θ. In this case an additional relationship between this force and the motion of the system must be established, and some of the advantage of Lagrange's equations is lost.

Problems

For Problems 17-21 through 17-40: (a) write expressions for the components of generalized force corresponding to the indicated coordinates, and (b) write expressions for the potential function for those forces that are conservative.

17-21 (a) Let the coordinates for the system of Problem 17-1 be x_1 and x_2, the horizontal displacements of the two particles, measured from an equilibrium configuration.
(b) Let the coordinates for the system of Problem 17-1 be x, locating the center of mass, and y, where the distance between particles is $l + y$ and the relaxed length of the spring is l.

17-22 (a) Let the coordinates for the system of Problem 17-2 be x_1 and x_2, the horizontal distances to the particles, measured from the equilibrium configuration.
(b) Let the coordinates for the system of Problem 17-2 be θ_1 and θ_2, the inclinations of the rods from the vertical.

17-23 For the system of Problem 17-3, let β be the angle between the left-hand rod and the horizontal, where the spring is relaxed. Let $\phi(t)$ be the inclination of the left-hand rod from the horizontal, for an arbitrary configuration of the system.

17-24 Let l be the distance from the pivot to the particle in the equilibrium configuration of the system of Problem 17-4. Let $l + x$ be this distance in an arbitrary configuration, and θ the inclination from the vertical.

17-25 Let x be the extension of the spring in the system of Problem 17-5 and θ the inclination of the pendulum from the vertical.

* See Section 5-3.

17-26   For the system of Problem 17-6, use the coordinates in Problem 17-25.

17-27   For the system of Problem 17-7, let θ be the inclination of the pendulum from the vertical and ϕ be the inclination of the line from the center of the flywheel to the driving pin, from the horizontal. (a) Let x be the extension of the spring. (b) Let y be the horizontal displacement of M from the position where $\phi = 0$ and the spring is unstretched.

17-28   For the system of Problem 17-8, let ϕ be the inclination of the pendulum from the radial line from the center of the flywheel to the pendulum support. For (b), let θ be the angle of rotation of the flywheel.

17-29   For the system of Problem 17-9, let ϕ be the inclination of the rod OP from the vertical.

17-30   For the system of Problem 17-10: (a) let x_1 and x_2 be the displacements of the left-hand and right-hand particles from their equilibrium positions, respectively; (b) let coordinates r_1 and r_2 be defined by

$$x_1 = 2r_1 + 2r_2$$
$$x_2 = (1 + \sqrt{5})r_1 + (1 - \sqrt{5})r_2$$

17-31   For the system of Problem 17-11, use the coordinates ϕ and χ as indicated in the diagram.

17-32   For the system of Problem 17-12, use the coordinate ϕ indicated in the diagram.

17-33   There is negligible friction between the block and the bottom of the groove in the system of Problem 17-13. The coefficient of friction between the block and the sides of the broove is μ. Let x be the radial distance from the center of rotation (where the spring forces balance) to the block.

17-34   In the flyball governor mechanism of Problem 17-14, let χ be the inclination of each arm from the vertical, and let ϕ be the angle of rotation around the vertical, of the projection of an arm onto a horizontal plane. The spring force is zero when $\chi = 0$.

17-35   In the system of Problem 17-15, let ϕ be the inclination of the pendulum from the vertical.

17-36   In the system of Problem 17-16, the bearing constrains the thin circular disc to remain perpendicular to the lower arm of the T bar. The fixed bearings in which the T bar rotates are in a horozontal line. Use the coordinates χ and ψ indicated in the diagram.

17-37   In the system of Problem 17-17, let ϕ be the angle between AC and the vertical.

17-38   In the system of Problem 17-18, let y be the vertical displacement of M

from its position in the equilibrium configuration, and let ϕ be the inclination of the pendulum from the vertical.

17-39   In the system of Problem 17-19, let s be the distance along the rod to the sliding particle from its position in the equilibrium configuration, and let ϕ be the angle from the vertical to the rod.

17-40   In the system of Problem 16-38, use the coordinates indicated.

17-3

LAGRANGE'S EQUATIONS.   Inertia forces may be included along with interaction forces in the principle of virtual work by replacing $\mathbf{f}_i$ with $\mathbf{f}_i - m_i\ddot{\mathbf{r}}_i$ in Equation 7-3. This leads to the relationship

$$\sum_{k=1}^{p} (m_k\ddot{\mathbf{r}}_k - \mathbf{f}_k)\cdot\delta\mathbf{r}_k = 0 \qquad (17\text{-}5a)$$

which is often called *D'Alembert's* principle.

A more convenient form may be obtained by expressing the virtual work of the inertia forces in a form analogous to Equation 17-2 for the interaction forces. Expressing the virtual displacement as before, we have

$$\sum_{k=1}^{p} m_k\ddot{\mathbf{r}}_k\cdot\delta\mathbf{r}_k = \sum_{k=1}^{p} m_k\ddot{\mathbf{r}}_k\cdot\left(\sum_{i=1}^{n} \frac{\partial\mathbf{r}_k}{\partial q_i}\,\delta q_i\right)$$

$$= \sum_{i=1}^{n}\left(\sum_{k=1}^{p} m_k\ddot{\mathbf{r}}_k\cdot\frac{\partial\mathbf{r}_k}{\partial q_i}\right)\delta q_i \qquad (a)$$

Next, the coefficient in this expansion may be rewritten as

$$\sum_{k=1}^{p} m_k\ddot{\mathbf{r}}_k\cdot\frac{\partial\mathbf{r}_k}{\partial q_i} = \sum_{k=1}^{p}\left[\frac{d}{dt}\left(m_k\dot{\mathbf{r}}_k\cdot\frac{\partial\mathbf{r}_k}{\partial q_i}\right) - m_k\dot{\mathbf{r}}_k\cdot\frac{d}{dt}\left(\frac{\partial\mathbf{r}_k}{\partial q_i}\right)\right] \qquad (b)$$

But, from Equation 17-1,

$$\frac{d}{dt}\left(\frac{\partial\mathbf{r}_k}{\partial q_i}\right) = \sum_{j=1}^{n} \frac{\partial}{\partial q_j}\left(\frac{\partial\mathbf{r}_k}{\partial q_i}\right)\dot{q}_j + \frac{\partial}{\partial t}\left(\frac{\partial\mathbf{r}_k}{\partial q_i}\right)$$

and

$$\frac{\partial}{\partial q_i}(\dot{\mathbf{r}}_k) = \frac{\partial}{\partial q_i}\left(\sum_{j=1}^{n} \frac{\partial\mathbf{r}_k}{\partial q_j}\dot{q}_j + \frac{\partial\mathbf{r}_k}{\partial t}\right)$$

Now the $2n + 1$ variables $(q_1, q_2, \cdots, q_n, \dot{q}_1, \dot{q}_2, \cdots, \dot{q}_n, t)$ will be considered as independent insofar as the meaning of partial differentiation is concerned. Then it follows from the preceding two equations that

$$\frac{d}{dt}\left(\frac{\partial \mathbf{r}_k}{\partial q_i}\right) = \frac{\partial \mathbf{v}_k}{\partial q_i} \tag{c}$$

where

$$\mathbf{v}_k = \dot{\mathbf{r}}_k \tag{d}$$

Also, from Equation 17-1, we have

$$\mathbf{v}_k = \sum_{j=1}^{n} \frac{\partial \mathbf{r}_k}{\partial q_j} \dot{q}_j + \frac{\partial \mathbf{r}_k}{\partial t}$$

from which

$$\frac{\partial \mathbf{r}_k}{\partial q_i} = \frac{\partial \mathbf{v}_k}{\partial \dot{q}_i} \tag{e}$$

Substitution of Equations c, d, and e into b leads to

$$\sum_{k=1}^{p} m_k \ddot{\mathbf{r}}_k \cdot \frac{\partial \mathbf{r}_k}{\partial q_i} = \sum_{k=1}^{p} \left[\frac{d}{dt}\left(m_k \mathbf{v}_k \cdot \frac{\partial \mathbf{v}_k}{\partial \dot{q}_i}\right) - m_k \mathbf{v}_k \cdot \left(\frac{\partial \mathbf{v}_k}{\partial q_i}\right)\right]$$

$$= \frac{d}{dt}\left(\frac{\partial}{\partial \dot{q}_i} \sum_{k=1}^{p} \tfrac{1}{2} m_k \mathbf{v}_k \cdot \mathbf{v}_k\right) - \frac{\partial}{\partial q_i} \sum_{k=1}^{p} \tfrac{1}{2} m_k \mathbf{v}_k \cdot \mathbf{v}_k$$

But the quantity $\sum_k \tfrac{1}{2} m_k v_k{}^2$ is the kinetic energy of the system. Denoting this by T, we can now write D'Alembert's principle as

$$\sum_{i=1}^{n} \left[\frac{d}{dt}\left(\frac{\partial T}{\partial \dot{q}_i}\right) - \frac{\partial T}{\partial q_i} - Q_i\right] \delta q_i = 0 \tag{17-5b}$$

Since this must be true for an arbitrary virtual displacement, the coefficient of each δq_i must vanish:

$$\boxed{\frac{d}{dt}\left(\frac{\partial T}{\partial \dot{q}_i}\right) - \frac{\partial T}{\partial \dot{q}_i} = Q_i} \tag{17-6}$$

These are the n *Lagrange's equations*, that govern the motion of the system.

Example

Figure 17-2 shows the velocities of the two mass particles in the system of Figure 17-1. Referring to this, we may express the kinetic energy of the system as

$$T = \tfrac{1}{2}m_1[\dot{s}^2 + (s\dot{\alpha})^2] + \tfrac{1}{2}m_2\{[\dot{s} + l\dot{\theta} \sin (\alpha - \theta)]^2 + [s\dot{\alpha} + l\dot{\theta} \cos (\alpha - \theta)]^2\}$$
$$= \tfrac{1}{2}[m\dot{\alpha}^2 s^2 + 2m_2 l\dot{\alpha}\dot{s} \cos (\alpha - \theta)\dot{\theta} + m\dot{s}^2 + 2m_2 l \sin (\alpha - \theta)\dot{s}\dot{\theta} + m_2 l^2\dot{\theta}^2]$$

where

$$m = m_1 + m_2$$

The inertia force terms in Lagrange's equations are then

$$\frac{d}{dt}\left(\frac{\partial T}{\partial \dot{s}}\right) - \frac{\partial T}{\partial s} = \frac{d}{dt}[m\dot{s} + m_2 l \sin (\alpha - \theta)\dot{\theta}]$$
$$- [m\dot{\alpha}^2 s + m_2 l\dot{\alpha} \cos (\alpha - \theta)\dot{\theta}]$$
$$= m(\ddot{s} - \dot{\alpha}^2 s) + m_2 l[\sin (\alpha - \theta)\ddot{\theta} - \cos (\alpha - \theta)\dot{\theta}^2]$$

and

$$\frac{d}{dt}\left(\frac{\partial T}{\partial \dot{\theta}}\right) - \frac{\partial T}{\partial \theta} = \frac{d}{dt}\{m_2 l[\dot{\alpha}s \cos (\alpha - \theta) + \sin (\alpha - \theta)\dot{s} + l\dot{\theta}]\}$$
$$- m_2 l[\dot{\alpha}s \sin (\alpha - \theta)\dot{\theta} - \cos (\alpha - \theta)\dot{s}\dot{\theta}]$$
$$= m_2 l[\ddot{s} - \dot{\alpha}^2 s) \sin (\alpha - \theta)$$
$$+ (s\ddot{\alpha} + 2\dot{\alpha}\dot{s}) \cos (\alpha - \theta) + l\ddot{\theta}]$$

Substitution of these expressions, and those of the components of generalized force (Equations d, p. 143), yields the equations of motion for the system:

$$m(\ddot{s} - \dot{\alpha}^2 s) + m_2 l[\sin (\alpha - \theta)\ddot{\theta} - \cos (\alpha - \theta)\dot{\theta}^2] + ks = ks_0 + mg \cos \alpha$$
$$(\ddot{s} - \dot{\alpha}^2 s) \sin (\alpha - \theta) + (s\ddot{\alpha} + 2\dot{\alpha}\dot{s}) \cos (\alpha - \theta) + l\ddot{\theta} + g \sin \theta = 0$$

Figure 17-2

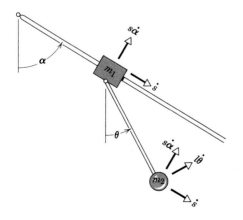

Problems

For Problems 17-41 through 17-60, write Lagrange's equations of motion in terms of the indicated coordinates.

17-(40+i) The system of Problem 17-(20+i) (i = 1, 2, · · · , 20).

17-4

INTEGRALS OF LAGRANGE'S EQUATIONS. Lagrange's equations provide an efficient procedure for deriving the equations of motion of complex mechanical systems. The next step in the investigation of a dynamic system—the integration of these equations—is often a formidable one, and may be possible to carry out only by approximations of some sort. General methods for obtaining exact solutions are known for some types of systems, such as the linear, constant coefficient systems to be discussed in Chapter 18. In this section, circumstances under which the differential order of certain systems can be reduced are pointed out. The first integrals that apply in these circumstances are generalizations of the conservation of mechanical energy and of momentum.

The Lagrangian Function. When the generalized force has no nonconservative components, so that all components are expressible as

$$Q_i = - \frac{\partial V}{\partial q_i}$$

it may be convenient to introduce the *Lagrangian function*, defined as

$$L(q_i, \dot{q}_i, t) = T(q_i, \dot{q}_i, t) - V(q_i, t) \tag{17-7}$$

Lagrange's equations can then be represented in terms of this function as

$$\frac{d}{dt} \left(\frac{\partial L}{\partial \dot{q}_i} \right) - \frac{\partial L}{\partial q_i} = 0 \tag{17-8}$$

The Hamiltonian Function. Consider the time derivative of the Lagrangian function,

$$\frac{dL}{dt} = \sum_{i=1}^{n} \frac{\partial L}{\partial q_i} \dot{q}_i + \sum_{i=1}^{n} \frac{\partial L}{\partial \dot{q}_i} \ddot{q}_i + \frac{\partial L}{\partial t}$$

With the introduction of Equation 17-8, this equation may be rewritten as

$$\sum_{i=1}^{n} \left[\frac{d}{dt} \left(\frac{\partial L}{\partial \dot{q}_i} \right) \dot{q}_i + \frac{\partial L}{\partial \dot{q}_i} \ddot{q}_i \right] - \frac{dL}{dt} + \frac{\partial L}{\partial t} = 0$$

or

$$\frac{dH}{dt} + \frac{\partial L}{\partial t} = 0 \qquad (17\text{-}9)$$

in which the quantity

$$H = \sum_{i=1}^{n} \frac{\partial L}{\partial \dot{q}_i} \dot{q}_i - L \qquad (17\text{-}10)$$

is called the *Hamiltonian function*. Its relationship with the energy of the system is brought out in the following development.

The expression for the kinetic energy can contain terms in which the generalized velocity components are absent, terms in which they appear linearly, and terms in which they appear quadratically. This becomes apparent on regrouping terms in the expression

$$T = \tfrac{1}{2} \sum_{i=1}^{p} m_k \mathbf{v}_k \cdot \mathbf{v}_k$$

$$= \tfrac{1}{2} \sum_{k=1}^{p} m_k \left(\sum_{i=1}^{n} \frac{\partial \mathbf{r}_k}{\partial q_i} \dot{q}_i + \frac{\partial \mathbf{r}_k}{\partial t} \right) \cdot \left(\sum_{j=1}^{n} \frac{\partial \mathbf{r}_k}{\partial q_i} \dot{q}_j + \frac{\partial \mathbf{r}_k}{\partial t} \right)$$

$$= \tfrac{1}{2} \sum_{k=1}^{p} m_k \frac{\partial \mathbf{r}_k}{\partial t} \cdot \frac{\partial \mathbf{r}_k}{\partial t} + \sum_{i=1}^{n} \left(\sum_{k=1}^{p} m_k \frac{\partial \mathbf{r}_k}{\partial t} \cdot \frac{\partial \mathbf{r}_k}{\partial q_i} \right) \dot{q}_i$$

$$+ \tfrac{1}{2} \sum_{i=1}^{n} \sum_{j=1}^{n} \left(\sum_{k=1}^{p} m_k \frac{\partial \mathbf{r}_k}{\partial q_i} \cdot \frac{\partial \mathbf{r}_k}{\partial q_j} \right) \dot{q}_i \dot{q}_j$$

which then has the form

$$T = T_0 + T_1 + T_2$$

where the subscripts indicate the degree of the functions of velocity components:

$$T_0 = k(q_i, t)$$

$$T_1 = \sum_{j=1}^{n} l_j(q_i, t) q_j$$

$$T_2 = \tfrac{1}{2} \sum_{j=1}^{n} \sum_{k=1}^{n} m_{jk}(q_i, t) \dot{q}_j \dot{q}_k$$

These forms may be readily identified in the expression for the kinetic energy of the slider-pendulum system discussed earlier.

Now the Hamiltonian may be written as

$$H = \sum_{i=1}^{n} \frac{\partial T_0}{\partial \dot{q}_i} \dot{q}_i + \sum_{i=1}^{n} \frac{\partial T_1}{\partial \dot{q}_i} \dot{q}_i + \sum_{i=1}^{n} \frac{\partial T_2}{\partial \dot{q}_i} \dot{q}_i - (T_0 + T_1 + T_2 - V)$$

Application of Euler's theorem on homogeneous functions reduces this to

$$H = 0 + T_1 + 2T_2 - (T_0 + T_1 + T_2 - V)$$
$$= T_2 + V - T_0 \tag{17-11}$$

Whenever t does not appear explicitly in the transformation (17-1), $T_0 = T_1 = 0$ and the Hamiltonian is equal to the total mechanical energy $T + V$.

An Energy Integral.   If the Lagrangian does not contain t explicitly, Equation 17-9 reduces to

$$\frac{dH}{dt} = 0$$

and we immediately have the first integral

$$H = H_0 \text{ (constant)} \tag{17-12}$$

An example is the special case $\alpha = $ constant, of the system shown in Figure 17-1. The energy integral for this case is

$$m\dot{s}^2 + 2m_2 l \sin (\alpha - \theta)\dot{s}\dot{\theta} + m_2 l^2 \dot{\theta}^2 + k(s - s_0)^2 - mg(s \cos \alpha + l \cos \theta) = H_0$$

Generalized Momentum.   The quantity

$$p_i = \frac{\partial T}{\partial \dot{q}_i} \tag{17-13}$$

is called the ith component of the *generalized momentum* of the system. Note that in the case of the system of Figure 17-1,

$$p_s = m\dot{s} + m_2 l \sin (\alpha - \theta)\dot{\theta}$$

is the component parallel to the guide rod of the linear momentum of the system, and the component

$$p_\theta = m_2 l[\dot{\alpha} s \cos (\alpha - \theta) + \sin (\alpha - \theta)\dot{s} + l\dot{\theta}]$$

is the magnitude of the moment about the hinge of the linear momentum.

Ignorable Coordinates.   Occasionally a system and the choice of coordinates are such that one or more of the coordinates does not appear explicitly in the expression for kinetic energy and the corresponding generalized force component is zero. If the coordinate q_r is missing from T, and $Q_r = 0$, the rth equation of motion is

$$\frac{dp_r}{dt} = 0$$

which may be readily integrated to yield

$$p_r = \beta_r \qquad (17\text{-}14)$$

where β_r is a constant depending on the initial conditions. An example is the special case $k = 0$ and $\alpha = \pi/2$, of the system in Figure 17-1. For this case the Lagrangian

$$L = \tfrac{1}{2}(m\dot{s} + 2m_2l \cos \theta \, \dot{s}\dot{\theta} + m_2l^2\dot{\theta}^2) - m_2gl(1 - \cos \theta)$$

is a function of $\dot{s}$, θ, and $\dot{\theta}$, but not of s. Therefore, the first-order equation

$$p_s = m\dot{s} + m_2l \cos \theta \, \dot{\theta} = \beta_s$$

can replace the first of the two Lagrange equations written earlier. The order of the system of differential equations has been reduced from fourth to third.

The coordinates satisfying $\dot{p}_r = 0$ are called *ignorable coordinates*. This terminology results from the fact that it is possible to write a set of equations of motion for the system in which none of the ignorable coordinates appears. A procedure for elimination of the ignorable coordinates, due to E. J. Routh, is described in L. Meirovitch, *Methods of Analytical Dynamics*, McGraw-Hill, 1970.

The Spinning Top Again.   Let us analyze the spinning top of Figure 16-13 by means of Lagrangian methods. The kinetic and potential energies may be expressed as

$$T = \tfrac{1}{2}[I_1\dot{\chi}^2 + I_1(\dot{\phi} \sin \chi)^2 + I_s(\dot{\phi} \cos \chi + \dot{\psi})^2] \qquad (a)$$

$$V = mgl \cos \chi \qquad (b)$$

Observe that of the three coordinates ϕ, χ, and ψ, the only one that occurs explicitly in Equations a and b is χ. Therefore, both ψ and ϕ are ignorable, and the two corresponding equations of motion,

$$\dot{p}_\psi = \frac{d}{dt} \, [I_s(\dot{\psi} + \dot{\phi} \cos \chi) = 0 \qquad (c)$$

$$\dot{p}_\phi = \frac{d}{dt} \, [I_s \, \dot{\psi} \cos \chi + \dot{\phi}(I_s \cos^2 \chi + I_1 \sin^2 \chi)] = 0 \qquad (d)$$

may be integrated once:

$$I_s \dot{\psi} + I_s \cos \chi \dot{\phi} = \beta_\psi \tag{e}$$

$$I_s \cos \chi \dot{\psi} + (I_s \cos^2 \chi + I_1 \sin^2 \chi)\dot{\phi} = \beta_\phi \tag{f}$$

Two of the equations of motion derived earlier by considering moments of forces and momentum, are equivalent to the above equations for the ψ and ϕ components of generalized momentum. Equation 16-16c is identical to Equation c above. Equation d can be obtained by multiplying (16-16b) by $\sin \chi$, multiplying (16-16c) by $\cos \chi$, and adding the two resulting equations.

Now, if Equations e and f are used to eliminate $\dot{\psi}$ and $\dot{\phi}$ from Equation a, the conservation of mechanical energy can be expressed as

$$\dot{\chi}^2 + U(\chi) = \frac{2E - \beta_\psi^2/I_s}{I_1} = E_0 \tag{g}$$

where

$$U(\chi) = \left(\frac{\beta_\psi \cos \chi - \beta_\phi}{I_1 \sin \chi} \right)^2 + \frac{mgl}{I_1} \cos \chi \tag{h}$$

Note that this has the same mathematical structure as (11-3b). Following the procedure there, the variables may be separated, leading to

$$\pm \int \frac{d\chi}{\sqrt{E_0 - U(\chi)}} = t_0 + t \tag{i}$$

With the substitution $\cos \chi = u$, this integral may be transformed into an elliptic integral, tabulated values of which are available. Further details may be found in Whittaker's book. The analysis predicts oscillatory motion such as depicted in the sketch on p. 132.

Problems

For Problems 17-61 through 17-80, either indicate that no general integral of Lagrange's equation exists, or write out any integrals that do exist.

17-(60+i)   The system of Problem 17-(20+i). (i = 1, 2, $\cdots$, 20).

17-81   Two compartments of a rotating space station are modeled as two particles interconnected by the elastic and dissipative elements shown. The motion can be shown to take place in a plane. (See Problem 12-51.)
(a) In terms of the four coordinates x, y, ϕ, and z, write the Lagrange equations of motion.
(b) Write any integrals of these equations you can.
(c) Eliminate the ignorable coordinate(s) from the Lagrange equation(s) for the nonignorable coordinate(s).

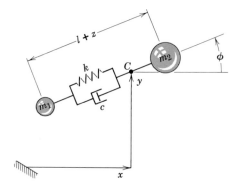

17-82   A top such as that shown in Figure 16-13 has the parameters

$$l = 10 \text{ mm}$$
$$m = 4.0 \text{ kg}$$
$$I_s = 4.0 \text{ g} \cdot \text{m}^2$$
$$I_1 = 4.5 \text{ g} \cdot \text{m}^2$$

It is set in motion with initial conditions

$$\dot{\psi} = 60 \text{ rad/s}$$
$$\chi = 45^0$$
$$\dot{\phi} = 1.80 \text{ rad/s}$$

(a) Evaluate the total mechanical energy $T + V$, and the components p_ϕ and p_ψ of generalized momentum.

(b) Determine the limits of χ between which oscillation will take place. *Suggestion.* Note that at extreme values of χ, $\dot{\chi}$ will vanish.

(c) Estimate the frequency of oscillation. *Suggestion.* Determine where $dU/d\chi$ vanishes and expand $U(\chi)$ in its Taylor series about this point, up through the second-degree term. Using this approximation, compare Equation g with the conservation of mechanical enegy for the simple spring-mass system, for which the frequency of oscillation is $\omega = \sqrt{k/m}$.

DYNAMIC
BEHAVIOR
OF SYSTEMS

Previously, we have emphasized the basic laws of mechanics and the correct mathematical expression of these laws. Sometimes these expressions directly give answers of engineering usefulness. Often, however, the result is a set of differential equations that, although they properly govern the motion, require considerable further analysis before a sufficient understanding of the system is achieved. Such analysis may be relatively straightforward (as in the case of the particle falling in a gravitational field), or the extraction of desired information may be difficult and expensive.

In this chapter we examine the methods of integration that work for a widely encountered class of dynamic systems known as *linear, constant-coefficient* systems. Motions of systems that fall outside this class are generally more difficult to predict, and no thorough examination of appropriate methods will be undertaken here. However, the interesting behavior of a few *nonlinear* systems will be presented.

18-1

A SAMPLING OF PHYSICAL SYSTEMS. The physical systems presented in this section illustrate the various procedures of predicting and inter-

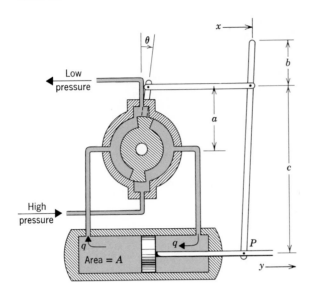

Figure 18-1

preting dynamic behavior. Initially the analysis of each is carried through the *modeling* stage, that is, the translation of the knowledge of physical characteristics into equations that may be expected to predict the system behavior with acceptable accuracy. In later sections we take up the integration of equations of motion and the interpretation of the solutions.

A Hydraulic Servoactuator. Figure 18-1 shows the essential features of a device that employs pressurized fluid to give a "power assist" to the positioning of an aircraft control surface. The device operates as follows: if the positions of the control stick and airfoil are such that the valves are open, fluid will flow, causing a repositioning of the airfoil.

If the openings are small, the forces from fluid pressure are likely to predominate over inertia forces; therefore, in a first approximation, let us neglect the mass in the moving parts. Let us assume further that the flow rate is a known function of the valve opening:

$$\text{Flow rate} \qquad q = f(\theta) \qquad\qquad (a)$$

This function will undoubtedly depend on the geometry of the opening, properties of the fluid, and the pressure drop between chambers. A typical form of this is shown in Figure 18-2. Since the fluid is essentially incompressible, the flow rate must be equal to the rate at which the piston sweeps out cylinder volume; that is,

$$q = -A \frac{dy}{dt} \qquad\qquad (b)$$

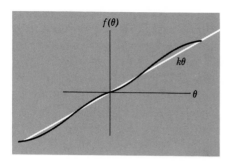

Figure 18-2

Note the consistency of the minus sign with the definitions for positive y and q, Figure 18-1. The remaining element of the description of the system follows from consideration of the linkage connecting the control stick position x, the valve opening θ, and the airfoil position y. As the solution to Problem 18-1 shows, the quantities are related by

$$a\theta = \frac{by + cx}{b + c} \tag{c}$$

The above three equations govern the three variables y, θ, and q. With a little algebra, the last two of these may be eliminated, yielding a single equation for y:

$$A\frac{dy}{dt} + f\left(\frac{by + cx}{ab + ac}\right) = 0 \tag{d}$$

Solution of this differential equation may be relatively easy or difficult, depending on the form of the function $f(\theta)$. Perhaps the easiest to manage would be the *linear* form indicated by the dashed line in Figure 18-2, and represented mathematically by

$$f(\theta) = k\theta \tag{e}$$

This assumption allows further simplification of the governing equation to

$$\frac{dy}{dt} + By = Cx \tag{18-1}$$

in which the constants are related to quantities introduced earlier by

$$B = \frac{kb}{Aa(b + c)} \qquad C = -\frac{kc}{Aa(b + c)}$$

Equation 18-1 is of the same form as that which governs the particle falling in a viscous medium.[*] In this chapter we introduce an approach to its integration that is somewhat different from that outlined in Chapter 11.

[*] Charles Smith, *Dynamics*, Wiley, New York, 1976, pp. 64–65.

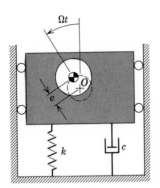

Figure 18-3

A Single-Degree-of-Freedom Oscillator.

Rotating machines typically possess a certain amount of unbalance, causing vibratory forces to be transmitted through the mounting. Figure 18-3 shows a simplified model of such a machine. Between the machine and the foundation are a spring with a stiffness constant k (force per unit deflection) and a *dashpot*, a device that transmits a force equal to c times the velocity between its ends. The total mass of the machine is denoted by m, that of the rotor by m_r, and the distance from the bearing centerline O to the mass center of the rotor by e. The assumptions that the rotor rotates at a specified rate and that the machine is constrained against all motion other than vertical result in a model with one degree of freedom.

Let us specify the vertical displacement of the machine with the coordinate y, measured upward from the position at which the spring transmits zero force. The forces acting on the system consisting of the machine (including the rotor) are shown in the freebody diagram of Figure 18-4. The sum of these forces must equal the rate of the change of momentum of the system. Referring to Figure 18-3, we find that the momentum of the nonrotating parts is equal to $(m - m_r) \dot{y}$ upward, and the upward component of the momentum of the rotor

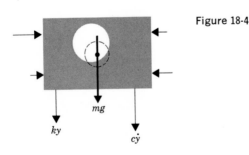

Figure 18-4

is equal to m_r $(\dot{y} - e\Omega \sin \Omega t)$. Thus, the equation governing the motion of the machine is

$$-ky - c\dot{y} - mg = \frac{d}{dt} [(m - m_r)\dot{y} + m_r(\dot{y} - e\Omega \sin \Omega t)]$$

Or, with a little rearrangement,

$$m\ddot{y} + c\dot{y} + ky = -mg + m_r e\Omega^2 \cos \Omega t \qquad (18\text{-}2)$$

This equation contains a surprising amount of useful information. We will extract and evaluate some of this after we present a method of integrating equations of this type.

An Electromechanical Shaker.   Figure 18-5 is a schematic of a device for vibration testing. It is designed to produce a displacement of the table proportional to the input electromotive force $E(t)$, which can be programmed electronically. The coil attached to the moving table is under the influence of a strong, steady magnetic field. A change in the current through this coil causes a corresponding change in the force transmitted from the steady field and, therefore, a corresponding change in the table's position of equilibrium with the support springs.

Figure 18-5

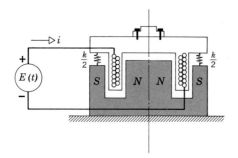

Consider first the interactions that take place as the current-carrying conductor moves through the magnetic field. Motion of the coil will induce an emf along the conductor, proportional to the velocity v of the coil*:

$$e_m = Kv \qquad (a)$$

The constant K is called the electromagnetic coupling constant. Its value depends on the intensity and geometry of the magnetic field and the geometry

---

* The phenomenon of electromechanical coupling is discussed in detail in texts on electrical fundamentals. See, for instance, Halliday and Resnick, *Fundamentals of Physics*, Wiley, 1970.

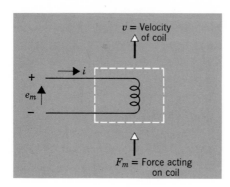

Figure 18-6

of the coil, and can be positive or negative depending on the polarity of the field and the sign conventions for emf and velocity. Current passing through the coil will induce a force on the coil proportional to the current i:

$$F_m = Ki \tag{b}$$

The constant in this equation is the same as that in Equation a. The signs in the two equations are consistent with the sign conventions indicated in Figure 18-6; that is, current i is taken as positive flowing in the direction of decreasing emf e_m, and directions of positive force and positive velocity are the same. An understanding of the equivalence of the two equations can be gained by working Problems 18-2 and 18-3.

Kirchoff's law for potential may be stated as

$$E(t) = L\frac{di}{dt} + Ri + e_m \tag{c}$$

in which R denotes the resistance in the loop and L the self-inductance. Application of Newton's second law to the free body shown in Figure 18-7 leads to

$$m\frac{d^2y}{dt^2} = -mg - kx + F_m \tag{d}$$

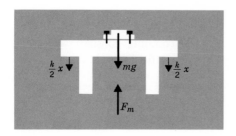

Figure 18-7

where y is the displacement measured upward from the position at which the spring reaction is zero. Introducing the relationships (a) and (b) into (c) and (d) yields the following equations governing current and displacement:

$$L \frac{di}{dt} + Ri + K \frac{dy}{dt} = E(t)$$

$$m \frac{d^2y}{dt^2} + ky - Ki = -mg$$

(18-3)

After a method of integrating equations of this type is presented we will examine how well the displacement y actually *does* follow some input signals.

A Two-Degree-of-Freedom Structure. The cantilever beam shown in Figure 18-8 carries two particles that can move in the vertical plane as the beam flexes. Assuming that the particles are constrained to move vertically, and that the beam itself has negligible mass, the system is one with two degrees of freedom. Thus, to carry out an analysis, we need two coordinates, such as y_1 and y_2, to specify an arbitrary configuration of the system. Load-deflection relationships for beams are given in the many texts on mechanics of deformable structures*; for the cantilever in this example the reactions shown in Figure 18-9 are related to the deflections by

$$\frac{96EI}{7a^3} y_1 - \frac{30EI}{7a^3} y_2 = f_1$$

$$-\frac{30EI}{7a^3} y_1 + \frac{12EI}{7a^3} y_2 = f_2$$

(a)

in which EI is the *flexural rigidity* of the beam, depending on its material and cross section. Newton's second law for the particles is expressed by

$$m_1 g - f_1 = m_1 \ddot{y}_1$$

$$m_2 g - f_2 = m_2 \ddot{y}_2$$

(b)

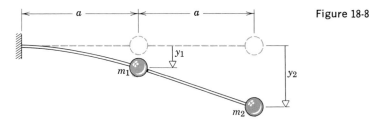

Figure 18-8

* See, for instance, Crandall and Dahl, *An Introduction to the Mechanics of Solids*, McGraw-Hill, 1972.

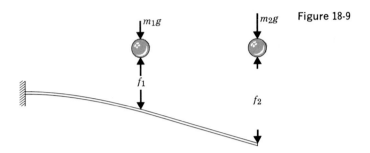

Figure 18-9

Elimination of f_1 and f_2 from (a) and (b) results in the following two equations that govern the displacements.

$$m_1\ddot{y}_1 + \frac{96EI}{7a^3}y_1 - \frac{30EI}{7a^3}y_2 = m_1g$$

$$m_2\ddot{y}_2 - \frac{30EI}{7a^3}y_1 + \frac{12EI}{7a^3}y_2 = m_2g \qquad (18\text{-}4)$$

In Section 5-4 we examine the vibratory motion that these equations predict.

An Infinite-Degree-of-Freedom Structure.   The cantilever beam model depicted in Figure 18-10 embodies the concept of mass distributed continuously along its length. To develop useful equations from this idea, we introduce a set of variables that specify the lateral displacement of each mass point along the beam. Thus, in analogy with $y_1(t)$ and $y_2(t)$ of the previous example, we consider the set of displacements $y(\xi,t)$, where each value of ξ between 0 and L specifies a particular mass point by giving its location along the ξ axis.

When the beam is in motion, each element transmits an inertia force proportional to the local acceleration. To relate this to the elastic restoring force, we once again borrow from structural analysis the relationships between elastic deflection and a distributed load (per unit length) $f(\xi)$. These are

$$(EIy'')'' = f(\xi) \qquad\qquad 0 < \xi < L \qquad\qquad \text{(a)}$$

$$y = y' = 0 \qquad\qquad \xi = 0 \qquad\qquad (18\text{-}5a)$$

$$EIy'' = (EIy'')' = 0 \qquad\qquad \xi = L \qquad\qquad (18\text{-}5b)$$

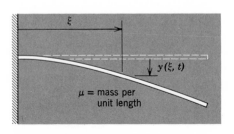

Figure 18-10

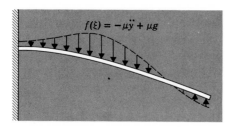

Figure 18-11

where the prime indicates differentiation with respect to ξ. The dynamic state is then accounted for by introducing the inertia and gravity forces as the applied load, as indicated in Figure 18-11.

$$f(\xi) = -\mu \frac{\partial^2 y}{\partial t^2} + \mu g \tag{b}$$

Then the motion is governed by Equations 18-5 above and the following combination of (a) and (b):

$$\mu \ddot{y} + (EIy'')'' = \mu g \qquad 0 < \xi < L \tag{18-5c}$$

The motions that Equations 18-5 predict are similar to those predicted by Equations 18-4.

Recapitulation. The five examples just introduced were selected so that a representative variety of commonly encountered aspects of modeling could be illustrated. For this reason the examples may appear to have little in common beyond their dealing with time-varying quantities. This initial view is misleading, however, as will be brought out in subsequent sections, where it will become apparent that the similarities among these systems outweigh their differences.

Before continuing, you should review the examples, with particular attention to the following aspects of the modeling procedure, which must always be considered.

First, decisions must be made concerning which effects of a real system can be reasonably neglected, so that the resulting model is not unnecessarily complex. (An example is the decision to neglect inertia forces in the hydraulic servoactuator.) Each of the examples contains a number of such simplifying assumptions on which the governing equations rest; some of the problems are intended to call your attention to these, and provoke critical thought about the limitations they impose on the adequacy of the resulting predictions.

Second, having conceived a simplified model in physical terms, equations of motion that are consistent with the physical model must be properly constructed. To accomplish this, we first introduce *variables* or quantities that we expect to vary with time, almost always in an unknown way, and that will play a part in the basic laws that govern the components of the system. The positive sense and the origin for each of these variables must be clearly defined.

Next, with the system considered in *an arbitrary state of motion*, we write an equation expressing each physical law that the various components must obey, in terms of the variables introduced. This done, if there are as many independent* equations as variables, we can turn to mathematical operations. If there are more variables than independent equations, a law has been overlooked. For example, Equations a, b, and c, pp. 160 and 161, contain the variables q, θ, and y. Since the three equations contain exactly three variables, it is reasonable to turn to mathematical operations, such as the combination of these to form the single equivalent Equation d.

The steps enumerated above should be identified for each of the examples of this section, and given careful attention as exercises are worked.

Problems

18-1   Derive the relationship (c), p. 161. Make a sketch that clearly shows each step. What restrictions are there on the validity of (c)?

18-2   Show that K in Equation a, p. 163, is dimensionally consistent with K in Equation b, p. 164.

18-3   Consider the rate of energy transfer in the moving coil discussed on pp. 163–164. Transfer of electrical energy to other forms can be expressed in terms of potential drop of charge moving through the coil, as $P = e_m i$. Or, from the point of view of force that the field applies to the coil, as $P = F_m v$. Use these to verify the consistency of Equations a and b.

18-4   An object of mass m is suspended by the spring-dashpot arrangement shown. The system is excited through the prescribed vertical displacement of the support, $a(t)$. Derive the equation(s) of motion.

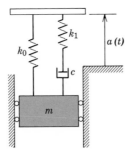

---

* By independent is meant that none of the equations can be deduced from others solely by mathematical operation; usually, intuition is reliable for avoiding derivation of redundant equations.

18-5 The accelerometer shown is forced to follow a given displacement $y(t)$. Write the differential equation governing the relative displacement $x(t)$.

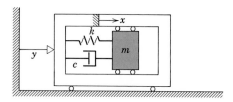

18-6 The pendulum-beam suspension is intended to isolate the instrument package from ground motions. Write the equations that govern θ and y.

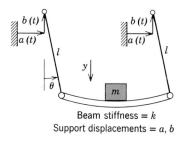

Beam stiffness = k
Support displacements = a, b

18-7 Two particles of equal mass are attached to the tightly stretched elastic line. The tension in a disturbed configuration differs little from the initial tension S_0. Derive the equations governing small vertical motions.

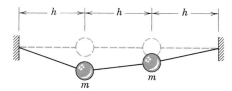

18-8 The left-hand end of the line of Problem 18-7 is given a vertical displacement $a(t)$. Derive the equations of motion for this case.

18-9 A tightly stretched elastic line has a tension S and mass per unit length μ. Derive the equations governing small lateral motions.

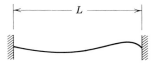

18-10 The railroad car couples with the bumper on striking it. Write equations of motion (both differential equations and initial conditions), and state clearly the idealizations implied in these.

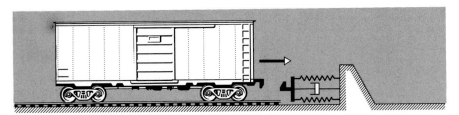

Although the virtual work idea developed by Lagrange and discussed in Chapter 17 makes many of the derivations of differential equations more direct, the force-acceleration approach also works. For those you have not already done, write the differential equations governing the idealized systems in the following problems.

18-11 Problem 17-5.

18-12 Problem 17-6.

18-13 Problem 17-10.

18-14 Problem 17-13.

18-15 Problem 17-17.

18-16 Problem 17-18.

18-17 Problem 17-81.

18-2

CLASSIFICATION OF SYSTEMS. This section is devoted to identifying important properties of systems, in terms of the mathematical form of the governing differential equations. Recognizing certain characteristics of equations of motion is essential in deciding how to approach integration. Furthermore, important aspects of system behavior can often be predicted before integration is undertaken.

Lumped and Distributed Parameters; Order of Systems. The entities that have values that can be assigned prior to integration of equations of motion will be considered as the *parameters* of the system; for example, A, a, b, c, and k in the hydraulic actuator of Section 18-2. Entities that have values to be determined through solution of the equations of motion are called the *dependent variables*, or simply the *variables* of the system; for example, q, θ, and y of the same actuator. In differential equations, those quantities with respect to which derivatives appear are considered the *independent variables*, for example, t in all the previous examples, and ξ and t in the last example.*

When the model contains a finite number of parameters, we call it a *lumped parameter* model. A *distributed parameter* model is one containing an infinite set of parameters. The first four systems of Section 18-1 were modeled with lumped parameters. In the last example, the model employs values of bending rigidity EI and mass density μ at all points along the beam; thus, we have here a distributed parameter model. Lumped parameter models are typically described by sets of *ordinary* differential equations, that is, equations in which derivatives with respect to only one independent variable appear. Distributed parameter models are typically described by *partial* differential equations, that is, equations in which derivatives with respect to two or more independent variables appear.

The order of a differential equation is the order of the highest derivative that appears. The order of a dynamic system governed by a set of ordinary differential equations is the sum of the orders of the governing equations. As modeled, the systems of the first four examples are, in the order of appearance, first order, second order, third order, and fourth order.

Linear and Nonlinear Systems.   If the only appearance of the dependent variables and their derivatives is as a linear combination in the equations of motion, the system is said to be *linear*. Otherwise, the system is *nonlinear*. The system defined by (d), p. 161 is nonlinear because y appears nonlinearly in the function f; the assumption (e) reduces it to the linear system defined by Equation 18.1. The models for the other examples in Section 18-1 are all linear.

This distinction is important because integration other than by approximate means is seldom possible for nonlinear systems. Although such approximate solutions are usually constructed readily with the aid of computers, they typically require extensive interpretation of many curves in order that a thorough understanding of the system can be achieved. A linear approximation, although limited in the range of its validity, can usually provide insight into the properties of the system with comparatively little interpretation of the solution. For this reason we normally consider modeling the real physical system as linear whenever this approximation can be justified.

The parameters defining a linear system appear in the mathematical description as the coefficients of the dependent variables and their derivatives.

---

* Although important, the distinction between independent and dependent variables is a matter of prejudice. For example, the differential equation $(d^2y/dt^2) + ay = 0$ may be thought of as defining the dependent variable y as a function of the independent variable t, $y = f(t)$; but an alternate form of the same equation, $(d^2t/dy^2) - ay(dt/dy)^3 = 0$ could be thought of equally well as defining the dependent variable t as a function of the independent variable y, $t = g(y)$.

When these are all constant with respect to time, the system is called a *linear, constant-coefficient system*, or l.c.c. system. A linear system with time varying parameters would arise, for example, if the cantilever beam in Figure 18-8 were provided with a support that could slide along the beam in such a way as to make the stiffness parameters prescribed functions of time. Time-varying parameters also introduce difficulties against integration of the equations of motion.

Autonomous and Nonautonomous Systems.   If time t does not appear explicitly in the equations of motion, the system is called *autonomous*. If t does appear explicitly, the system is *nonautonomous*. When the appearance of t is in the form of time-varying coefficients of the dependent variables, the non-autonomous system is said to be *parametrically excited*. A function of t appearing as a term separate from the dependent variables is usually called a *forcing function*. The systems governed by Equations 18-1, 18-2, and 18-3 are non-autonomous, excited by the forcing functions $Cx(t)$, $m_r e \Omega^2 \cos \Omega t$, and $E(t)$, respectively. The pendulum under the influence of the moving support,. Problem 17-17, is a parametrically excited nonautonomous system. The systems governed by Equations 18-4 and 18-5 are autonomous.

Various Equivalent Forms of Equations of Motion.   Through straightforward manipulation it is possible to cast equations of motion into a number of equivalent forms. Since discussions in the literature are carried out with reference to some "standard" forms, it is important to readily recognize the equivalence of different structures of equations.

A standard form for an nth-order system, sometimes called a *state-variable form*, consists of a set of n first order equations containing n variables:

$$\frac{d\, y_j}{dt} = f_j(y_1, y_2, \cdots y_n, t) \qquad j = 1, 2, \cdots, n \qquad (18\text{-}6)$$

In the case of a linear system, this would be specialized as

$$\frac{d\, y_j}{dt} = \sum_{k=1}^{n} a_{jk}\, y_k + F_j(t) \qquad (18\text{-}7)$$

or appear in the matrix equivalent

$$\frac{d}{dt}\, \{y\} = [a]\{y\} + \{F\} \qquad (18\text{-}8)$$

If modeling results in equations of order higher than first, the higher-order equations can be replaced with several equations of first order. For illustration, consider the third-order system governed by Equations 18-3, p. 165. We first write (merely as a definition of the variable v)

$$\frac{dy}{dt} = v$$

then introduce this into Equations 18-3, with the result

$$\frac{di}{dt} = -\frac{R}{L} i - \frac{K}{L} v + E(t)$$

$$\frac{dv}{dt} = \frac{K}{m} i - \frac{k}{m} y - g$$

These three equations constitute a state variable form of the system description. In matrix form, it would appear as

$$\frac{d}{dt} \left\{ \begin{array}{c} i \\ v \\ y \end{array} \right\} = \left[\begin{array}{ccc} -\dfrac{R}{L} & -\dfrac{K}{L} & 0 \\ \dfrac{K}{m} & 0 & -\dfrac{k}{m} \\ 0 & 1 & 0 \end{array} \right] \left\{ \begin{array}{c} i \\ v \\ y \end{array} \right\} + \left\{ \begin{array}{c} E(t) \\ -g \\ 0 \end{array} \right\}$$

At another extreme, we could also reduce the number of equations in the set to one, by eliminating the variable i from Equation 18-3, with the result

$$\frac{d^3 y}{dt^3} + \frac{R}{L} \frac{d^2 y}{dt^2} + \left(\frac{k}{m} + \frac{K^2}{mL} \right) \frac{dy}{dt} + \frac{kR}{mL} y = \frac{KE(t)}{mL} - \frac{Rg}{L}$$

Any of the above forms is suitable for the classification of the system, and, as we will see, for the integration to obtain the system response.

In structural dynamics, modeling usually results in equations of the form

$$[m]\{\ddot{q}\} + [c]\{\dot{q}\} + [k]\{q\} = \{Q(t)\} \tag{18-9}$$

in which the elements in q are displacement coordinates, representing the deformed configuration of the structure. For example, Equations 18-4 have this form:

$$\left[\begin{array}{cc} m_1 & 0 \\ 0 & m_2 \end{array} \right] \left\{ \begin{array}{c} \ddot{y}_1 \\ \ddot{y}_2 \end{array} \right\} + \frac{6EI}{7a^3} \left[\begin{array}{cc} 16 & -5 \\ -5 & 2 \end{array} \right] \left\{ \begin{array}{c} y_1 \\ y_2 \end{array} \right\} = \left\{ \begin{array}{c} m_1 g \\ m_2 g \end{array} \right\}$$

These may also be readily cast into the state variable form, but there is seldom any advantage to this.

Problems

Classify each of the following.

18-18 (a) $\dfrac{d^2y}{dt^2} - 2\dfrac{dy}{dt} + 7y = 0$

(b) $\dfrac{\partial^2 u}{\partial x^2} + y^2 u = 0$

(c) $\dfrac{\partial^2 u}{\partial x^2} + a^2\dfrac{\partial^2 u}{\partial y^2} = 0$

(d) $\dfrac{d^2y}{dt^2} + \omega^2 y = \cos \Omega t$

(e) $\dfrac{d^2y}{dt^2} + (a + b\cos \Omega t)y = 0$

(f) $\dfrac{d^2y}{dt^2} + y\dfrac{dy}{dt} + 2y = 0$

18-19 (a) $3\dfrac{d^2x}{dt^2} + \dfrac{d^2y}{dt^2} + 2x - y = 0$

$\dfrac{d^2x}{dt^2} + 2\dfrac{d^2y}{dt^2} - x + 3y = 0$

(b) $3\dfrac{d^2x}{dt^2} + \dfrac{d^2y}{dt^2} + 2x - y = 3\cos \Omega t$

$\dfrac{d^2x}{dt^2} + 2\dfrac{d^2y}{dt^2} - x + 3y = -\cos \Omega t$

(c) $3\dfrac{d^2x}{dt^2} + \dfrac{d^2y}{dt^2} + (2 - \cos \Omega t)x - y = 0$

$\dfrac{d^2x}{dt^2} + 2\dfrac{d^2y}{dt^2} - x + (3 + \tfrac{1}{2}\cos \Omega t)y = 0$

18-20 Equations 16-16.

18-21 (a) $EI\dfrac{\partial^4 u}{\partial x^4} + \mu\dfrac{\partial^2 u}{\partial t^2} = 0$

(b) $EI\dfrac{\partial^4 u}{\partial x^4} + \mu\dfrac{\partial^2 u}{\partial t^2} = w(x,t)$

18-22 (a) $\dfrac{\partial^2 \psi}{\partial t^2} - \dfrac{E}{\rho}\dfrac{\partial^2 \psi}{\partial x^2} - \dfrac{kG}{\rho r^2}\left(\dfrac{\partial u}{\partial x} - \psi\right) = 0$

$\dfrac{\partial^2 u}{\partial t^2} - \dfrac{kG}{\rho}\left(\dfrac{\partial^2 u}{\partial x^2} - \dfrac{\partial \psi}{\partial x}\right) = 0$

(b) $\dfrac{\rho}{kG}\dfrac{\partial^4 u}{\partial t^4} - \left(1 + \dfrac{E}{kG}\right)\dfrac{\partial^4 u}{\partial x^2 \partial t^2} + \dfrac{E}{\rho}\dfrac{\partial^4 u}{\partial x^4} + \dfrac{1}{r^2}\dfrac{\partial^2 u}{\partial t^2} = 0$

18-23 For every equation in Section 18-1, list which quantities are parameters and which quantities are variables.

18-(23+i) Classify each system model you derived among Problems 18-(3+i) (i = 1, 2,···, 14).

18-38 Show the equivalence between the state variable form for a linear second-order system,

$$\left\{\begin{array}{c} \dot{y}_1 \\ \dot{y}_2 \end{array}\right\} - \left[\begin{array}{cc} a_{11} & a_{12} \\ a_{21} & a_{22} \end{array}\right]\left\{\begin{array}{c} y_1 \\ y_2 \end{array}\right\} = \left\{\begin{array}{c} F_1(t) \\ F_2(t) \end{array}\right\}$$

and the equation of motion (18-20); that is, determine the values for ζ, ω_0, and $F(t)$ in terms of the quantities in the state variable description. Construct the state variable form (18-8) from the equation of motion (18-20).

18-3

LINEARIZATION. All real systems contain some nonlinearities. In most cases, if the ranges of values of the dependent variables are sufficiently restricted, the system may be well approximated as linear. In many cases, the variables fortunately never vary outside these ranges, and a linear analysis will suffice for the complete study. And, even when the system is expected to operate in the nonlinear ranges, a linear analysis can still provide useful information. Therefore, rational procedures for making the linear approximation to a nonlinear system are very important.

The simplest way to arrive at a linear model is by assuming linearity in all laws governing components of a system; examples are the assumed forms of (c) and (e), p. 161. However, many times it is not easy to obtain a good estimate of the proportionality constants. When this is the case, we may include nonlinearities in the model, and then determine the proper proportionality constant through a mathematical procedure that we now illustrate.

Application of Newton's law of motion to the idealized mechanism shown in Figure 18-12 leads to the nonlinear differential equation

$$(ml^2 + 4Ml^2 \sin^2 \phi)\ddot{\phi} + 2Ml^2 \sin 2\phi \, \dot{\phi}^2 - 4kl^2 \sin \phi(\cos \phi - \cos \beta)$$
$$+ \, mgl \cos \phi = 0 \qquad \text{(a)}$$

For restricted ranges of the angle ϕ, a linear approximation to the nonlinear functions in the equation would be satisfactory. However, unless the values of ϕ are near an equilibrium value, the system can be expected to soon travel outside the range in which the approximation is valid. For this reason, we normally are interested in the linearization *near an equilibrium point*. Therefore, the first step is to determine equilibrium value(s) ϕ_0. This may be accomplished by specializing the differential equation to describe the equilibrium state, that is, by setting all time derivatives to zero:

$$-4kl^2 \sin \phi_0(\cos \phi_0 - \cos \beta) + mgl \cos \phi_0 = 0 \qquad \text{(b)}$$

Next, we define a new variable θ with its origin at equilibrium,

$$\phi(t) = \phi_0 + \theta(t) \qquad \text{(c)}$$

and develop the Taylor's series for the nonlinear functions of the variables and their derivatives:

$$f(\theta,\ddot{\theta}) = [ml^2 + 4Ml^2 \sin^2(\phi_0 + \theta)]\ddot{\theta}$$
$$= [(ml^2 + 4Ml^2 \sin^2 \phi_0 + (4Ml^2 \sin 2\phi_0)\theta + \cdots]\ddot{\theta}$$

$$g(\theta,\dot{\theta}) = 2Ml^2 \sin 2(\phi_0 + \theta)\dot{\theta}^2$$
$$= [2Ml^2 \sin 2\phi_0 + (4Ml^2 \cos 2\phi_0)\theta + \cdots]\dot{\theta}^2$$

$$h(\theta) = -4kl^2 \sin(\phi_0 + \theta)[\cos (\phi_0 + \theta) - \cos \beta)] + mgl \cos (\phi_0 + \theta)$$
$$= [-4kl^2 \sin \phi_0 (\cos \phi_0 - \cos \beta) + mgl \cos \phi_0]$$
$$+ \{-mgl \sin \phi_0 + 4kl^2[\sin^2 \phi_0 - \cos \phi_0(\cos \phi_0 - \cos \beta)]\}\theta$$
$$+ \cdots$$

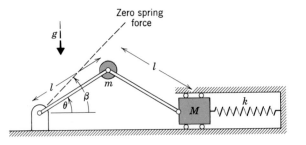

Zero spring force

Figure 18-12

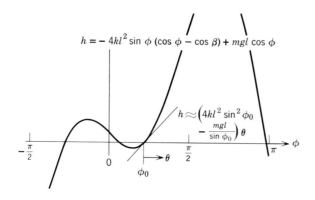

Figure 18-13

$$h = -4kl^2 \sin \phi (\cos \phi - \cos \beta) + mgl \cos \phi$$

$$h \approx \left(4kl^2 \sin^2 \phi_0 - \frac{mgl}{\sin \phi_0}\right)\theta$$

Introduction of (b) into the last expression reduces it to

$$h(\theta) = \left(4kl^2 \sin^2 \phi_0 - \frac{mgl}{\sin \phi_0}\right)\theta + \cdots$$

Linearization is then accomplished by dropping all powers higher than the first of θ and its derivatives, resulting in

$$(ml^2 + 4ml^2 \sin^2 \phi_0)\ddot{\theta} + \left(4kl^2 \sin^2 \phi_0 - \frac{mgl}{\sin \phi_0}\right)\theta = 0 \qquad (d)$$

The meaning of this process may perhaps be better understood from Figure 18-13. The proportionality constant K in the approximate linear law, $h(\theta) = K\theta$ is taken as the local slope of the nonlinear function h at the equilibrium point.

It is interesting to note from the figure that there are four equilibrium points for this example, and the proportionality constant is negative for two of them and positive for the other two. This sign has important ramifications for the motion of the system; we will return to this point later.

Another example to which the linearizing procedure can be applied is the torque-free rotation of a rigid body about its mass center. Euler's equations governing the three components of angular velocity along the principal axes of inertia are:

$$I_1\dot{\omega}_1 = (I_2 - I_3)\omega_2\omega_3$$
$$I_2\dot{\omega}_2 = (I_3 - I_1)\omega_3\omega_1 \qquad (a)$$
$$I_3\dot{\omega}_3 = (I_1 - I_2)\omega_1\omega_2$$

As you may verify, an "equilibrium" state that satisfies these equations is that in which the resultant spin is about the principal axis x_3:

$$\omega_1 = \omega_2 = 0 \qquad \omega_3 = \Omega$$

To study the behavior of the system in the neighborhood of this equilibrium state, we rewrite the differential equations in terms of the new variables η_i, which measure the deviation from the equilibrium:

$$\omega_1 = \eta_1$$
$$\omega_2 = \eta_2$$
$$\omega_3 = \Omega + \eta_3$$

Substitution into the above differential equations gives the description in terms of the η_i:

$$\dot{\eta}_1 = \frac{I_2 - I_3}{I_1}(\Omega\eta_2 + \eta_2\eta_3)$$

$$\dot{\eta}_2 = \frac{I_3 - I_1}{I_2}(\Omega\eta_1 + \eta_3\eta_1) \tag{c}$$

$$\dot{\eta}_3 = \frac{I_1 - I_2}{I_3}\eta_1\eta_2$$

Now, if the values of η_i are sufficiently small, it would seem a reasonably good approximation to neglect the products $\eta_i\eta_j$, leaving the linearized model

$$\dot{\eta}_1 = \frac{I_2 - I_3}{I_1}\Omega\eta_2$$

$$\dot{\eta}_2 = \frac{I_3 - I_1}{I_2}\Omega\eta_1 \tag{d}$$

$$\dot{\eta}_3 = 0$$

We shall return to this example after methods for integrating linear, constant coefficient systems are examined.

The general nonlinear, autonomous system is linearized about an equilibrium state as follows. First, the equilibrium point $(y_{10}, y_{20}, \cdots, y_{n0})$ is determined as a special solution of the equations of motion,

$$\dot{y}_j = f_j(y_1, y_2, \cdots, y_n) \qquad j = 1, 2, \cdots, n \tag{18-10}$$

in which the time derivatives are zero:

$$0 = f_j(y_{10}, y_{20}, \cdots, y_{n0})$$

Once a set of values y_{10} satisfying these equations is determined, new variables η_j, measuring excursions from the equilibrium point, are defined by

$$y_j(t) = y_{j0} + \eta_j(t) \tag{18-11}$$

These definitions are then substituted, and power series are developed for the functions f_j: *

$$\dot{\eta}_j = f_j(y_{10} + \eta_1, y_{20} + \eta_2, \cdots, y_{n0} + \eta_n)$$

$$= f_j(y_{10}, y_{20}, \cdots, y_{n0}) + \sum_{k=1}^{n} \left(\frac{\partial f_j}{\partial y_k}\right)_0 \eta_k$$

$$+ \frac{1}{2} \sum_{k=1}^{n} \sum_{l=1}^{n} \left(\frac{\partial^2 f_j}{\partial y_k \partial y_l}\right)_0 \eta_k \eta_l + \cdots$$

(If, as in the preceding example, the functions f_j are in the form of polynomials, the power series result directly from substitution.) Because the y_{j0} satisfy equilibrium, the first term is zero. Neglecting powers of η_j higher than first results in the linearized model:

$$\dot{\eta}_j = \sum_{k=1}^{n} a_{jk} \eta_k \qquad (18\text{-}12)$$

in which the coefficients are the slopes evaluated at the equilibrium point:

$$a_{jk} = \left(\frac{\partial f_j}{\partial y_k}\right)_0 \qquad (18\text{-}13)$$

Although a state variable form of the equations of motion was used as a basis for delineating the method, transformation to this form is not necessary. The essential steps, which may be taken using any form of governing equations, are:

1. Determination of the equilibrium point.
2. Definition of variables that measure excursions from the equilibrium point.
3. Development of all functions of the variables and their derivatives in power series.
4. Neglect of powers higher than first.

A similar procedure can be used to linearize a nonautonomous system. However, in this case, we cannot expect an equilibrium solution to exist; therefore, Step 1, above, is replaced with a determination of a solution $y_{j0}(t)$, which is normally a difficult task. Linearized equations can then be determined in terms of excursions about this solution, $\eta_j(t) = y_j(t) - y_{j0}(t)$, by the same procedure as for the autonomous system. The result in this case is a set of linear, *variable coefficient* equations.

---

* For Taylor's series for functions of several variables, see, for instance, G. A. Korn and T. M. Korn, *Mathematical Handbook for Scientists and Engineers*, McGraw-Hill, 1968.

Problems

18-39 Determine the equilibrium configuration for the system shown in Figure 18-8. Write equations of motion in terms of new variables η_i defined by $y_i(t) = y_{i0} + \eta_i(t)$.

18-40 Determine the equilibrium configuration $y_0(\xi)$ for the beam shown in Figure 18-10. Write the equations of motion in terms of a new set of variables $\eta(\xi,t)$ defined by

$$y(\xi,t) = y_0(\xi) + \eta(\xi,t)$$

18-41 (a) Derive the equation of motion for the device shown in Figure 18-12.
(b) Write an equivalent set of state variable equations.
(c) Following the procedure as outlined in Equations 18-10 through 18-13, linearize the system.

18-42 A particle of mass m is suspended at midspan on the elastic line. As the line stretches, its tension varies according to the linear law, $S = S_0 + k\,\Delta L$, where ΔL is the increase in length of the line from its straight configuration. Derive the equation governing lateral motion of the particle. Neglecting gravity, linearize the equation of motion.

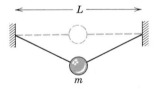

Linearize the equations modeling the systems of the following problems:

18-43 Problems 17-5, 18-11.

18-44 Problems 17-6, 18-12.

18-45 Problems 17-17, 18-15.

18-46 Problems 17-18, 18-16.

18-47 Problem 16-38.

18-4

IMPLICATIONS OF LINEARITY: SUPERPOSITION. Linear systems possess some properties of which an analyst can ill afford to lose sight. Because of their overriding importance, and the fact that these properties are easily demonstrated without recourse to the solutions to the equations of motion, special attention is given them in this section, before we become occupied with the details of solutions themselves.

By a *solution* to the set of differential equations (18-6) (or one of the equivalent forms) is meant a set of functions $y_1(t), y_2(t) \cdots y_n(t)$ that will satisfy these equations throughout a time interval. The differential equations possess many solutions, a *particular* solution being determined by a complete set of *initial conditions*, that is, a set of values for each state variable specified at the initial instant $t = 0.$*

For the discussions here, let us adopt the convention of placing on the left-hand sides of equations the terms containing the independent variables, and on the right-hand sides the forcing functions in the differential equations and the given initial values in the equations of initial conditions:

$$\{\dot{y}\} - [a]\{y\} = \{F(t)\}$$
$$\{y(0)\} = \{A\} \tag{18-14}$$

The following *theorem* may be readily verified by direct substitution into the equations of motion and initial conditions. *If the solution to*

$$\{\dot{y}\} - [a]\{y\} = \{F_1(t)\}$$
$$\{y(0)\} = \{A_1\}$$

is $y_1(t)$ and the solution to

$$\{\dot{y}\} - [a]\{y\} = \{F_2(t)\}$$
$$\{y(0)\} = \{A_2\}$$

is $y_2(t)$ then the solution to

$$\{\dot{y}\} - [a]\{y\} = \alpha\{F_1(t)\} + \beta\{F_2(t)\}$$
$$\{y(0)\} = \alpha\{A_1\} + \beta\{A_2\}$$

is $\alpha y_1(t) + \beta y_2(t)$. The same theorem may be stated in terms of a description of the system made up of second order equations (See Problem 18-49). This property gives us some help in construction of solutions to new problems in terms of solutions that have been already constructed.

For illustration, suppose we have somehow determined that the solution to the equations

$$\dot{y} - ay = 1 \qquad t \geq 0$$
$$y(0) = 0$$

---

* Occasionally there arises a peculiar circumstance in which more than one solution will satisfy the differential equation and a set of initial conditions. Here we assume the solution is unique. For a discussion of uniqueness see, for instance, E. L. Ince, *Ordinary Differential Equations*, Dover, 1956.

is

$$y_1(t) = \frac{e^{at} - 1}{a}$$

and that the solution to the equations

$$\dot{y} - ay = 0 \qquad t \geq 0$$
$$y(0) = 1$$

is $y_2(t) = e^{at}$. This information can be used to construct the solution to the equations

$$\dot{y} - ay = 5$$
$$y(0) = -2$$

because the pair of quantities in the right-hand sides, $\begin{Bmatrix} 5 \\ -2 \end{Bmatrix}$, can be made up of a combination of the two pairs of right-hand sides of the equation to which we have solutions:

$$\begin{Bmatrix} 5 \\ -2 \end{Bmatrix} = \alpha \begin{Bmatrix} 1 \\ 0 \end{Bmatrix} + \beta \begin{Bmatrix} 0 \\ 1 \end{Bmatrix}$$

By inspection, $\alpha = 5$ and $\beta = -2$, so that the solution we seek is

$$y = 5 \frac{e^{at} - 1}{a} - 2e^{at}$$

In a later section we will see how to extend this idea to construct solutions for completely arbitrary forcing functions, in terms of some "fundamental" solutions.

The equations represented by

$$\{\dot{y}\} - [a]\{y\} = \{0\} \tag{18-15}$$

which may be considered as a special case of Equation 18-14, are called the *auxiliary equations*, or *reduced equations*. Considering these as one of the sets of equations in the superposition theorem, p. 181, we find that *the sum of any solution to the reduced equations and a particular solution to the equations*

$$\{\dot{y}\} - [a]\{y\} = \{F(t)\} \tag{[18-14]}$$

is another particular solution to Equation 18-14.

For example, a particular solution to

$$\begin{Bmatrix} \dot{y}_1 \\ \dot{y}_2 \end{Bmatrix} - \begin{bmatrix} 0 & -1 \\ 1 & 0 \end{bmatrix} \begin{Bmatrix} y_1 \\ y_2 \end{Bmatrix} = \begin{Bmatrix} \sin 3t \\ 0 \end{Bmatrix} \tag{a}$$

is

$$\begin{Bmatrix} y_1 \\ y_2 \end{Bmatrix} = \begin{Bmatrix} -\frac{3}{8} \cos 3t \\ -\frac{1}{8} \sin 3t \end{Bmatrix} \qquad\qquad (b)$$

and a particular solution to the reduced equations is

$$\begin{Bmatrix} y_1 \\ y_2 \end{Bmatrix} = \begin{Bmatrix} \cos t \\ \sin t \end{Bmatrix}$$

Therefore, another particular solution to Equations a is

$$\begin{Bmatrix} y_1 \\ y_2 \end{Bmatrix} = \begin{Bmatrix} -\frac{3}{8} \cos 3t \\ -\frac{1}{8} \sin 3t \end{Bmatrix} + C_1 \begin{Bmatrix} \cos t \\ \sin t \end{Bmatrix}$$

where C_1 is any constant. It can also be verified that

$$\begin{Bmatrix} y_1 \\ y_2 \end{Bmatrix} = \begin{Bmatrix} -\sin t \\ \cos t \end{Bmatrix}$$

is a solution to the reduced equations, so that still another particular solution to Equations a is

$$\begin{Bmatrix} y_1 \\ y_2 \end{Bmatrix} = \begin{Bmatrix} -\frac{3}{8} \cos 3t \\ -\frac{1}{8} \sin 3t \end{Bmatrix} + C_1 \begin{Bmatrix} \cos t \\ \sin t \end{Bmatrix} + C_2 \begin{Bmatrix} -\sin t \\ \cos t \end{Bmatrix} \qquad (c)$$

In general, if $\{y\}_1, \{y\}_2, \cdots, \{y\}_n$ are linearly independent* solutions of the reduced equations (18-15), where n is the order of the system, then any solution to Equations 18-15 can be formed as a linear combination of these:

$$\{y\}_c = C_1\{y\}_1 + C_2\{y\}_2 + \cdots + C_n\{y\}_n$$

This combination, the general solution to the reduced equation, is called the *complementary function* for the system (18-14).

Furthermore, the most general solution to

$$\{y\} - [a]\{y\} = \{F(t)\} \qquad\qquad [18\text{-}14]$$

can be shown to be†

$$\{y\} = \{y\}_p + \{y\}_c$$

---

* By lineraly independent is meant that no one of the set of $\{y\}$s can be formed as a linear combination of the other members of the set. It can be shown that the $\{y\}$s are linearly independent if and only if the determinant $|\{y\}_1\{y\}_2 \ldots \{y\}_n| \neq 0$.
† See, for instance, E. L. Ince, *Ordinary Differential Equations*, Dover, 1956.

where $\{y\}_p$ is any particular solution and $\{y\}_C$ is the complementary function. Thus, for example, since

$$\begin{Bmatrix} \cos t \\ \sin t \end{Bmatrix} \quad \begin{Bmatrix} -\sin t \\ \cos t \end{Bmatrix}$$

are linearly independent, (c) is the *general solution* to the second-order system (a).

Section 18-5 is devoted to methods of finding the complementary function and particular solutions, and to evaluating the arbitrary constants in the general solution in terms of initial conditions.

Problems

18-48   A computer program has been developed that integrates any equation of the form

$$\frac{d^2y}{dx^2} + a(x)\,\frac{dy}{dx} + b(x)y = f(x)$$

where the information supplied to the computer consists of values of y and y' at $x = 0$, and sets of values of $a(x)$, $b(x)$, and $f(x)$ throughout the region $0 \leq x \leq L$. We wish to use this to solve the following problem

$$\frac{d^2y}{dx^2} + a(x)\,\frac{dy}{dx} + b(x)y = f(x) \qquad 0 \leq x \leq L$$

$$y(0) = 4$$
$$y(L) = 2$$

where we have all the information to input to the computer except $y'(0)$. Explain how we can make three runs with different combinations of $f(x)$, $y(0)$, and $y'(0)$, and explain how we can use the results to construct the desired solution.

18-49   (a) State and prove the superposition theorem for linear systems as applied to the form normally encountered in structural dynamics:

$$[m]\{\ddot{q}\} + [c]\{\dot{q}\} + [k]\,\{q\} = \{Q(t)\}$$
$$\{q(0)\} = \{A\}$$
$$\{q(0)\} = \{B\}$$

(b) State and prove the superposition theorem for linear systems as applied to the form of a single nth-order differential equation and the complete set of initial conditions:

$$a_n \frac{d^n y}{dt^n} + a_{n-1} \frac{d^{n-1} y}{dt^{n-1}} + \cdots + a_0 y = F(t)$$

$$y(0) = A$$
$$\dot{y}(0) = B$$
$$\cdot$$
$$\cdot$$
$$\cdot$$
$$y^{(n-1)}(0) = P$$

18-5

RESPONSE OF LINEAR, CONSTANT PARAMETER SYSTEMS.   This class of systems is the one for which an approach to analysis has been quite thoroughly developed, and can be counted on to yield answers. Since this is generally not the case for systems falling outside this classification, it is fortunate that a great many of the systems of interest to engineers can be well approximated as linear with constant parameters.

"Natural" Motions.   When the system is free from outside influences after an initial time, the motions are referred to as *natural motions*. Mathematically, this means that the motions are governed by equations in which the forcing function is zero, that is, the *reduced equations*. Therefore, the natural motions are those given by the complementary function.

First-order systems.   The first-order, linear differential equation may always be arranged in the form

$$\dot{y} - ay = F(t) \tag{18-16}$$

and the reduced equation that governs natural motion is then

$$\dot{y} - ay = 0 \tag{18-17}$$

A method of solving this (called separation of variables) was outlined in an earlier chapter. A perhaps less-straightforward approach, but one that will extend readily to higher-order systems, is motivated by the following observation. A function that is to satisfy Equation 18-17 must be one that retains its original form upon differentiation, that is, its derivative must be equal to the constant a times the function itself. This suggests trying $y = e^{\lambda t}$; and substitution into the left-hand side of Equation 18-17 leads to

$$\dot{y} - ay = (\lambda - a)e^{\lambda t}$$

If, as Equation 18-17 requires, this is to be zero for all t, we must have

$$\lambda - a = 0 \tag{18-18}$$

This is called the *characteristic equation* for the system, and gives us the value of λ, the *characteristic root*. The substitution above also verifies that with the proper choice of λ, $e^{\lambda t}$ is indeed a solution to Equation 18-17. You can readily verify that

$$y = C_1 e^{at} \tag{18-19}$$

in which C_1 is an arbitrary constant, is also a solution to Equation 18-17. The number of linearly independent solutions found being equal to the order of the system, this expression is the *general solution* to Equation 18-17, or the *complementary function* for Equation 18-16.

This result gives some information about the behavior of the hydraulic controller of Figure 18-1. If the input displacement x is held at zero, the displacement y will be governed by the equation reduced from (18-1):

$$\frac{dy}{dt} + By = 0$$

Comparison with Equation 18-17 reveals that the value for a in Equation 18-17 is in this case

$$a = -B$$

so that the output displacement varies according to

$$y(t) = C_1 e^{-Bt}$$

Thus, with the stick held at the zero position, an initially disturbed piston will return to close the valve, as is evident from Figure 18-1.

Second-order systems. The approach of the previous section works without modification if the second-order system appears in the form of a single differential equation,

$$\ddot{y} + 2\zeta\omega_0\dot{y} + \omega_0^2 y = F(t) \tag{18-20}$$

The apparently awkward nomenclature for the coefficients is introduced for later convenience in expressing solutions. The physical parameters in any second-order system can be determined in terms of these by comparison of the differential equation with Equation 18-20. For example, those of the system of Figure 18-3 can be determined by comparison of Equations 18-2 and 18-20:

$$\omega_0^2 = \frac{k}{m} \qquad 2\zeta\omega_0 = \frac{c}{m}$$

Or,

$$\omega_0 = \sqrt{\frac{k}{m}} \qquad \zeta = \frac{c}{2\sqrt{mk}}$$

As before, we recognize that a solution of the reduced equation must be such that its form is not altered on successive differentiation in order that the linear combination of $\ddot{y}$, $\dot{y}$, and y be zero for all time. In the present case, the substitution of $y = e^{\lambda t}$ into the reduced equation leads to

$$(\lambda^2 + 2\zeta\omega_0\lambda + \omega_0^2)e^{\lambda t} = 0$$

and, accounting for the fact that this must be satisfied for all t, we have

$$\lambda^2 + 2\zeta\omega_0\lambda + \omega_0^2 = 0 \qquad (18\text{-}21)$$

This is the characteristic equation for this system. It has the two roots given by the quadratic formula,

$$\left\{ \begin{matrix} \lambda_1 \\ \\ \lambda_2 \end{matrix} \right\} = \omega_0(-\zeta \pm \sqrt{\zeta^2 - 1}) \qquad (18\text{-}22)$$

Thus, there are two solutions that satisfy the reduced equation, $y = e^{\lambda_1 t}$, and $y = e^{\lambda_2 t}$. Except for a special circumstance, which we will examine later, these two solutions are linearly independent; therefore, according to information in the previous section, the general solution to the reduced equation (i.e., the complementary function) can be expressed as

$$y = C_1 e^{\lambda_1 t} + C_2 e^{\lambda_2 t} \qquad (18\text{-}23)$$

in which λ_1 and λ_2 are given by Equation 18-22.

The constants of integration can be determined in terms of the initial values of y and $\dot{y}$ as follows. Denoting these initial values as y_0 and v_0, and setting $t = 0$ in Equation 18-23 and its derivative, lead to

$$y(0) = C_1 + C_2 = y_0$$
$$\dot{y}(0) = \lambda_1 C_1 + \lambda_2 C_2 = v_0$$

Algebraic inversion of these equations yields

$$C_1 = \frac{-\lambda_2 y_0 + v_0}{\lambda_1 - \lambda_2} \qquad (18\text{-}24)$$

$$C_2 = \frac{\lambda_1 y_0 - v_0}{\lambda_1 - \lambda_2}$$

Damping ratio.   The parameter ζ is called the *damping ratio*, and, as will become evident shortly, has a significant effect on the natural motion of the system.

From Equation 18-22 we see that for $|\zeta| < 1$, the characteristic roots are complex. With the abbreviations

$$\alpha = -\zeta\omega_0$$
$$\omega = \sqrt{1 - \zeta^2}\,\omega_0 \qquad (18\text{-}25)$$

the roots may be written as

$$\lambda_1 = \alpha + i\omega$$
$$\lambda_2 = \alpha - i\omega = \lambda_1{}^* \qquad (18\text{-}26)$$

In this case, if the solution (18-23) is to be real, the constants of integration must be complex conjugates of one another, i.e., $C_2 = C_1{}^*$. (See Appendix E.) With the polar representation of the integration constants,

$$C_1 = \frac{C}{2}\,e^{i\theta} \qquad C_2 = \frac{C}{2}\,e^{-i\theta} \qquad (18\text{-}27)$$

the solution may be written as

$$y = \frac{C}{2}\,e^{i\theta}e^{(\alpha+i\omega)t} + \frac{C}{2}\,e^{-i\theta}e^{(\alpha-i\omega)t}$$

$$= Ce^{\alpha t}\,\frac{e^{i(\omega t+\theta)} + e^{-i(\omega t+\theta)}}{2}$$

$$= Ce^{\alpha t}\cos(\omega t + \theta) \qquad (18\text{-}28)$$

Thus, the motion in this case is oscillatory. The real part of the characteristic roots, $\alpha = -\zeta\omega_0$, gives the rate of decay (or growth, in the case of negative ζ) of the amplitude of motion, and the imaginary part, $\omega = \sqrt{1 - \zeta^2}\,\omega_0$, gives the frequency of oscillation.

The constants of integration C and θ depend on the initial conditions. To determine this dependence we can insert Equations 18-24 and 18-26 into Equation 18-27, with the result

$$y_0 + i\,\frac{\alpha y_0 - v_0}{\omega} = Ce^{i\theta}$$

From the relationships between rectangular and polar forms (see Appendix E) follow

$$C\cos\theta = y_0$$

$$C\sin\theta = \frac{\alpha y_0 - v_0}{\omega}$$

Insertion of these into the expanded form of Equation 18-28 gives the motion in terms of the given initial values y_0 and v_0:

$$y = e^{\alpha t}(C \cos \theta \cos \omega t - C \sin \theta \sin \omega t)$$

$$= e^{\alpha t}\left(y_0 \cos \omega t + \frac{v_0 - \alpha y_0}{\omega} \sin \omega t\right) \qquad (18\text{-}29)$$

For the case $|\zeta| > 1$, it is convenient to express the solution in terms of the parameter

$$\beta = \sqrt{\zeta^2 - 1} \; \omega_0 = i\omega \qquad (18\text{-}30)$$

Then,

$$\lambda_1 = \alpha + \beta \qquad \lambda_2 = \alpha - \beta$$

$$C_1 = \frac{-(\alpha - \beta)y_0 + v_0}{2\beta} \qquad C_2 = \frac{(\alpha + \beta)y_0 - v_0}{2\beta}$$

and the solution becomes

$$y = C_1 e^{(\alpha + \beta)t} + C_2 e^{(\alpha - \beta)t}$$

$$= e^{\alpha t}[\tfrac{1}{2}(C_1 + C_2)(e^{\beta t} + e^{-\beta t}) + \tfrac{1}{2}(C_1 - C_2)(e^{\beta t} - e^{-\beta t})]$$

$$= e^{\alpha t}\left(y_0 \cosh \beta t + \frac{v_0 - \alpha y_0}{\beta} \sinh \beta t\right) \qquad (18\text{-}31)$$

In the case $|\zeta| = 1$, the two characteristic roots become equal, and we no longer have two linearly independent solutions of the form $e^{\lambda t}$. The solution for this case can be found by letting $\zeta \to 1$ in either Equation 18-29 or Equation 18-31, and is

$$y = e^{-\omega_0 t}[y_0 + (v_0 + \omega_0 y_0)t] \qquad (18\text{-}32)$$

Figure 18-14 shows the response resulting from the initial conditions $\dot{y}(0) = 0$, $y(0) = y_0$, for various values of the damping ratio. The damping that yields $\zeta = 1$ is called the *critical damping*. This value separates the *underdamped* systems, which exhibit oscillatory natural motion, from the *overdamped* systems, which exhibit exponential natural motions.

The response of the *undamped* system is given by Equation 18-29 with $\zeta = 0$, which then agrees with the result derived earlier.*

* Charles Smith, *Dynamics*, Wiley, New York, 1976, p. 67.

Figure 18-14

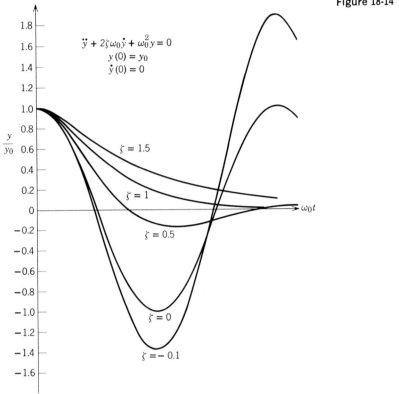

$$\ddot{y} + 2\zeta\omega_0\dot{y} + \omega_0^2 y = 0$$
$$y(0) = y_0$$
$$\dot{y}(0) = 0$$

A third-order system. Natural motions of the electromechanical shaker system of Figure 18-5 are governed by the equations reduced from (18-3):

$$L\frac{di}{dt} + Ri + K\frac{dy}{dt} = 0$$

$$m\frac{d^2y}{dt^2} + ky - Ki = 0$$

(18-33)

As in the previous two systems, we note that linear combinations of the variables and their derivatives are to equal zero; thus, we can expect that exponential variations in y and i will satisfy the equations of motion. However, we are now dealing with two equations, each containing both variables, and in this situation it has been found that in addition to varying exponentially with time, the variables must be in particular ratios to one another. That is, there are additional constants Y and I, appearing in the form

$$\left\{ \begin{matrix} i(t) \\ y(t) \end{matrix} \right\} = \left\{ \begin{matrix} I \\ Y \end{matrix} \right\} e^{\lambda t}$$

which must have particular values. Like λ, these are found following substitution into the differential equations. Substitution into Equations 18-33 leads to

$$\begin{bmatrix} (L\lambda + R) & K\lambda \\ -K & (m\lambda^2 + k) \end{bmatrix} \left\{ \begin{matrix} I \\ Y \end{matrix} \right\} = \left\{ \begin{matrix} 0 \\ 0 \end{matrix} \right\} \qquad (a)$$

Determination of the unknowns λ, I, and Y is an eigenvalue problem, almost exactly like that of determining the principal moments of inertia and principal axes for a rigid body. The relevant aspects of this problem are re-stated here:

The trivial solution, $I = Y = 0$, is of little interest, simply giving the equilibrium state of the system. A nontrivial solution exists if and only if the determinant of the above square matrix vanishes:

$$\begin{vmatrix} (L\lambda + R) & K\lambda \\ -K & (m\lambda^2 + k) \end{vmatrix} = 0$$

Expansion yields the characteristic equation for the system:

$$\lambda^3 + \frac{R}{L} \lambda^2 + \left(\frac{k}{m} + \frac{K^2}{mL} \right) \lambda + \frac{kR}{mL} = 0 \qquad (b)$$

To each of the three eigenvalues (or characteristic roots) satisfying this equation corresponds an eigenvector, determined by (a) with the eigenvalue substituted therein. With the eigenvalue so substituted, only one of the two equations is independent; this means that only the ratio of Y to I is determined, leaving a multiplicative constant arbitrary.

Calculations are somewhat less awkward if the characteristic root is dimensionless, and if the two variables are dimensionally identical. To achieve this, let us define

$$\mu = \frac{L}{R} \lambda \qquad X = \frac{L}{K} I$$

Then (a) and (b) may be rewritten as

$$\begin{bmatrix} (1 + \mu) & \mu \\ -\dfrac{K^2 L}{mR^2} & \mu^2 + \dfrac{kL^2}{mR^2} \end{bmatrix} \left\{ \begin{matrix} X \\ Y \end{matrix} \right\} = \left\{ \begin{matrix} 0 \\ 0 \end{matrix} \right\} \qquad (18\text{-}34)$$

$$\mu^3 + \mu^2 + \left(\frac{kL^2}{mR^2} + \frac{K^2 L}{mR^2} \right) \mu + \frac{kL^2}{mR^2} = 0 \qquad (18\text{-}35)$$

For illustration, consider the following values:

$$m = 4.5 \text{ kg}$$
$$k = 87\ 500 \text{ N/m}$$
$$K = 89. \text{ N/A}$$
$$R = 0.13 \ \Omega$$
$$L = 4.0 \times 10^{-5} \text{ H}$$

With these values, Equations 18-34 and 18-35 become

$$
\begin{bmatrix} 1 + \mu & \mu \\ -4.166\ 206 & \mu^2 + 0.001\ 841 \end{bmatrix} \begin{Bmatrix} X \\ Y \end{Bmatrix} = \begin{Bmatrix} 0 \\ 0 \end{Bmatrix}
\tag{c}
$$

$$\mu^3 + \mu^2 + 4.168\ 047\ \mu + 0.001\ 8409 = 0 \tag{d}$$

The roots of the characteristic equation are

$$\mu_1 = -0.000\ 442$$
$$\mu_{2,3} = -0.499\ 779 \pm 1.979\ 350i$$

Substitution of μ_1 into either of the two equations in (c) leads to

$$\left(\frac{Y}{X} \right)_1 = 2262.9$$

Similar substitution of $\mu_{2,3}$ leads to

$$\left(\frac{Y}{X} \right)_{2,3} = 1.000\ 05\ e^{\pm 2.646\ 729i}$$

From these values, we may write the general solution as

$$
\begin{Bmatrix} x(t) \\ y(t) \end{Bmatrix} = C_1 \begin{Bmatrix} 1 \\ 2262.9 \end{Bmatrix} e^{-0.000\ 442(Rt/L)}
$$

$$
+ C_2 \begin{Bmatrix} 1 \\ 1.000\ 05\quad e^{2.646\ 73i} \end{Bmatrix} e^{(-0.499\ 78 + 1.979\ 35i)(Rt/L)}
$$

$$
+ C_3 \begin{Bmatrix} 1 \\ 1.000\ 05\ e^{-2.646\ 73i} \end{Bmatrix} e^{(-0.499\ 78 - 1.979\ 35i)(Rt/L)}
$$

Now, if x and y are to be real, C_3 must be the complex conjugate of C_2. In terms of the polar form of these,

$$C_2 = \frac{D}{2}\, e^{i\theta}$$

$$C_3 = C_2^* = \frac{D}{2}\, e^{-i\theta}$$

the above solution becomes

$$\left\{ \begin{array}{c} x \\ y \end{array} \right\} = C_1 \left\{ \begin{array}{c} 1 \\ 2262.9 \end{array} \right\} e^{-0.000\,442(Rt/L)}$$

$$+ D \left\{ \begin{array}{c} \cos\left(1.979\,\dfrac{Rt}{L} + \theta\right) \\[2ex] 1.000\,05 \cos\left(1.979\,\dfrac{Rt}{L} + 2.647 + \theta\right) \end{array} \right\} e^{-0.4998(Rt/L)}$$

The constants of integration, C_1, D, and θ depend on the initial conditions. The remaining quantities in the solution are characteristic of the system itself, that is, they depend on the system parameters: The decay rates, $-0.000\,44(R/L)$ and $-0.4998(R/L)$ are the real parts of the characteristic roots, while the oscillation frequency, $1.979(R/L)$, is the imaginary part of the complex roots. This natural frequency has the value

$$f = \frac{1.979}{2\pi}\,\frac{R}{L} = 1024 \text{ cycles/sec (Hz)}$$

The relative amplitude and phase between displacement and current, $1.000\,05(L/K)$ and 2.647, are the absolute value and phase in the complex amplitude ration Y/X.

nth-Order systems.   Equations governing natural motions of an nth order, l.c.c. system can be put in the form

$$\{\dot{y}\} - [a]\{y\} = \{0\} \tag{18-36}$$

which we will use here to describe the procedure for determining natural motions. If the modeling results in equations of higher order, it is not necessary to rewrite them in a state variable form; the procedure described here applies equally well to any form, as exemplified by the preceding system.

Solutions to Equation 18-36 of the form

$$\{y(t)\} = \{Y\}e^{\lambda t}$$

can be found by substitution, which leads to

$$[\lambda[1] - [a]]\{Y\} = \{0\} \tag{18-37}$$

The characteristic equation results from the condition that there be a nontrivial $\{Y\}$ satisfying Equation 18-37:

$$|\lambda[1] - [a]| = 0 \tag{18-38}$$

Expansion of the determinant results in a polynomial equation of nth degree:

$$a_0\lambda^n + a_1\lambda^{n-1} + \cdots + a_{n-1}\lambda + a_n = 0 \tag{18-39}$$

To each distinct root λ_k corresponds a characteristic vector $\{Y\}_k$, determined from Equation 18-37. When the roots are all distinct*, $n - 1$ of the n equations in (18-37) are independent; this means that each $\{Y\}$ is determined to within an arbitrary multiplicative constant. The general solution to Equation 18-36 then has the form

$$\{y(t)\} = C_1\{Y\}_1 e^{\lambda_1 t} + C_2\{Y\}_2 e^{\lambda_2 t} + \cdots + C_n\{Y\}_n e^{\lambda_n t} \tag{18-40}$$

The constants of integration C_k may be determined from initial values of the variables in $\{y\}$ by solving the n equations in

$$\{y(0)\} = C_1\{Y\}_1 + C_2\{Y\}_2 + \cdots + C_n\{Y\}_n$$

A term $\{Y\}_k e^{\lambda_k t}$ in the solution (18-40) is easy to interpret when λ_k is real: all variables undergo exponential decay or growth at the same rate, and their relative amplitudes are determined by the elements in $\{Y\}_k$.

Complex roots. When the root λ_k is complex, its complex conjugate is another root:

$$\lambda_k = \alpha_k + i\omega_k$$

$$\lambda_{k+1} = \lambda_k^* = \alpha_k - i\omega_k$$

This follows from the fact that the coefficients in Equation 18-39 are real. Similarly, the complex conjugate of an eigenvector corresponding to λ_k is an eigenvector corresponding to λ_k^*, that is, if λ_k and $\{Y\}_k$ satisfy Equation 18-37, then λ_k^* and $\{Y^*\}_k$ also satisfy this equation. Then, if the solution to Equation 18-36 is to be real, the constants of integration C_k and C_{k+1} must be complex conjugates of one another. The motion resulting from combination of these two terms can be determined as follows: we first write the integration constants and the elements in the eigenvectors in polar form

$$C_k = \frac{D_k}{2} e^{i\theta_k}$$

---

* The unusual case of repeated roots requires special attention. The forms that solutions can take in these cases are outlined in L. Meirovitch, *Methods of Analytical Dynamics*, McGraw-Hill, 1970. An example is Equation 18-32.

$$C_{k+1} = C_k^* = \frac{D_k}{2} e^{-i\theta_k}$$

$$Y_{jk} = R_{jk} e^{i\phi_{jk}}$$

$$Y_{j(k+1)} = Y_{jk}^* = R_{jk} e^{-i\phi_{jk}}$$

The index $_j$ denotes the $_j$th element in the column $\{Y\}_k$. The contribution to the motion from these two solutions is then

$$y_{jk} = C_k Y_{jk} e^{\lambda_k t} + C_k^* Y_{jk}^* e^{\lambda_k^* t}$$

$$= \frac{D_k}{2} e^{i\theta_k} R_{jk} e^{i\phi_{jk}} e^{(\alpha_k + i\omega_k)t} + \frac{D_k}{2} e^{-i\theta_k} R_{jk} e^{-i\phi_{jk}} e^{(\alpha_k - i\omega_k)t}$$

$$= D_k e^{\alpha_k t} R_{jk} \frac{e^{i(\omega_k t + \phi_{jk} + \theta_k)} + e^{-i(\omega_k t + \phi_{jk} + \theta_k)}}{2}$$

$$= D_k e^{\alpha_k t} R_{jk} \cos(\omega_k t + \phi_{jk} + \theta_k)$$

Thus, the solution corresponding to the complex conjugate pair λ_k and λ_k^*, and $\{Y\}_k$ and $\{Y^*\}_k$ has the form

$$\begin{Bmatrix} y_1 \\ y_2 \\ \cdot \\ \cdot \\ \cdot \\ y_n \end{Bmatrix} = D_k e^{\alpha_k t} \begin{Bmatrix} R_{1k} \cos(\omega_k t + \phi_{1k} + \theta_k) \\ R_{2k} \cos(\omega_k t + \phi_{2k} + \theta_k) \\ \cdot \\ \cdot \\ \cdot \\ R_{nk} \cos(\omega_k t + \phi_{nk} + \theta_k) \end{Bmatrix} \qquad (18\text{-}41)$$

The constants of integration D_k and θ_k depend on the initial conditions. The remaining quantities depend on the system parameters: the real and imaginary parts of the characteristic roots are, respectively, the rate of exponential growth or decay α_k, and the frequency of oscillation ω_k. The relative amplitudes R_{jk} and relative phases ϕ_{jk} are the absolute values and phases of the complex elements of the eigenvectors $\{Y\}_k$.

Problems

18-50   Suppose the vertical rod of the control linkage in Figure 18-1 extended below point P, and the control x was effected from the lower end. What would be the effect on the natural motion of the system? Does this agree with the behavior you would predict from examination of the diagram?

18-51   Suppose the input displacement x is held fixed at a value x_0 in the hydraulic actuator of Figure 18-1. Determine the corresponding equilibrium value for y. Write the differential equation governing the variable $\eta(t) = y(t) - y_0$. Determine the solution and sketch the curve for y versus t for the case in which $y(0) = 0$.

18-52   Show that the third order differential equation, p. 173, is equivalent to Equations 18-3, p. 165. Show that the characteristic equation derived from this form is the same as that derived from Equation 18-3.

18-53   In terms of m, c, and k, what are the values of ζ and ω_0 (Equation 18-20) for the system of Figure 18-3? What are the values of the characteristic roots? For what ranges will natural motions be oscillatory?

18-54   Substitute each of the solutions (18-28), (18-29), and (18-31) into the equation reduced from (18-20), to verify their validity.

18-55   Verify that Equations 18-29 and 18-31 satisfy the initial conditions

$$y(0) = y_0 \qquad \dot{y}(0) = v_0$$

18-56   Derive (18-32) from (18-29) and from (18-31).

18-57   Investigate the nature of the natural motions governed by negative ω_0^2:

$$\ddot{y} + 2\zeta\alpha_0\dot{y} - \alpha_0^2 y = 0 \qquad \alpha_0 \text{ real}$$

What physical systems can you think of that would be modeled by an equation of this form?

18-58   Show that the lower equation in (18-34) gives the same result for Y/X as the upper equation when μ satisfies Equation 18-35.

18-59   Determine the natural motions of the system governed by Equation 18-4 with equal masses.

18-60   Determine the natural motions given by Equation d, p. 178. What type of motion occurs when
   (a) $I_3 > I_1, I_2$?
   (b) $I_3 < I_1, I_2$?
   (c) $I_1 < I_3 < I_2$ or $I_2 < I_3 < I_1$?

Compare the results with the implications of Figure 16-11b.

Determine the natural motions for the following systems:

18-61   The system of Problem 18-4, with $m = 2$ kg, $k_0 = 800$ N/m, $k_1 = 1200$ N/m, $c = 16$ N·s/m.

18-62   Problem 18-7.

18-63   Problem 18-43.

18-64   Problem 18-44.

18-65   Problem 18-45.

18-66   Problem 18-46.

18-67   Problem 18-47.

Forced Responses. *Construction of some particular solutions.* The simplest forcing function is a constant. A particular solution satisfying constant forcing functions is a set of constant values for all the variables. These values are readily determined by substitution into the differential equations. For example, a particular solution for the system (18-4) will be in the form

$$\left\{ \begin{matrix} y_1(t) \\ y_2(t) \end{matrix} \right\} = \left\{ \begin{matrix} y_{10} \\ y_{20} \end{matrix} \right\}$$

Substitution into the equations of motion gives

$$\begin{bmatrix} \dfrac{96EI}{7a^3} & -\dfrac{30EI}{7a^3} \\[2ex] -\dfrac{30EI}{7a^3} & \dfrac{12EI}{7a^3} \end{bmatrix} \left\{ \begin{matrix} y_{10} \\ y_{20} \end{matrix} \right\} = \left\{ \begin{matrix} m_1g \\ m_2g \end{matrix} \right\}$$

from which the particular solution may be found by algebraic inversion:

$$\left\{ \begin{matrix} y_{10} \\ y_{20} \end{matrix} \right\} = \begin{bmatrix} \dfrac{a^3}{3EI} & \dfrac{5a^3}{6EI} \\[2ex] \dfrac{5a^3}{6EI} & \dfrac{8a^3}{3EI} \end{bmatrix} \left\{ \begin{matrix} m_1g \\ m_2g \end{matrix} \right\}$$

These values represent the static equilibrium deflections due to the gravitational forces.

Another commonly encountered forcing function is sinusoidal in form. An example is the system illustrated in Figure 18-3 and governed by Equation 18-2. Keeping in mind the superposition principle and the information in the preceding paragraphs, we see that the general solution to Equation 18-2 can be completed provided we have a particular solution to

$$\ddot{x} + 2\zeta\omega_0\dot{x} + \omega_0^2 x = \cos \Omega t \qquad (18\text{-}42)$$

This equation has a particular solution in which x varies sinusoidally at the same frequency as the driving force, and with a phase difference; that is, it has the form

$$x(t) = X \cos (\Omega t - \phi) \qquad (18\text{-}43)$$

Substitution into the differential equation will verify that this function is indeed a particular solution, and will yield the values of the amplitude X and the phase lag ϕ. The ensuing algebra is handled most efficiently if related complex quantities are considered, as follows.

Consider, along with the above differential equation, the equation

$$\ddot{y} + 2\zeta\omega_0\dot{y} + \omega_0^2 y = \sin \Omega t$$

Multiplication of this equation by the unit imaginary i, and addition to the above equation, lead to

$$\ddot{z} + 2\zeta\omega_0\dot{z} + \omega_0^2 z = e^{i\Omega t}$$

in which $z = x + iy$. Now the desired solution x will be the real part of the solution z, which has the form

$$z(t) = Ze^{i\Omega t}$$

Substitution into the governing differential equation gives

$$[(i\Omega)^2 + 2\zeta\omega_0(i\Omega) + \omega_0^2]Ze^{i\Omega t} = e^{i\Omega t}$$

which will be satisfied for all t provided that

$$Z = \frac{1}{\omega_0^2 - \Omega^2 + 2i\zeta\omega_0\Omega}$$

In terms of the polar form,

$$Z = X e^{-i\phi}$$

$$X = \frac{1}{\sqrt{(\omega_0^2 - \Omega^2)^2 + (2\zeta\omega_0\Omega)^2}} \tag{18-44}$$

$$\phi = \tan^{-1}\frac{2\zeta\omega_0\Omega}{\omega_0^2 - \Omega^2} \tag{18-45}$$

the solution is

$$z(t) = X e^{i(\Omega t - \phi)}$$

Thus the desired solution, $x = Re\ z$, is, as stated earlier,

$$x(t) = X \cos (\Omega t - \phi) \tag{18-43}$$

with the amplitude and phase given by Equations 18-44 and 18-45.

This procedure readily yields a particular solution for a sinusoidally excited system described by a set of simultaneous differential equations, such as

$$\{\ddot{y}\} - [a]\{y\} = \{F_0\} \cos \Omega t \tag{18-46}$$

A particular solution of the form

$$\{y(t)\} = Re\ \{Ze^{i\Omega t}\}$$

may be expected, and the values of the constants in $\{Z\}$ found by substitution of

$$\{z\} = \{Z\}\ e^{i\Omega t}$$

into the differential equations

$$\{\dot{z}\} - [a]\{z\} = \{F_0\}\, e^{i\Omega t}$$

This substitution leads to

$$[i\Omega[1] - [a]]\{Z\} = \{F_0\}$$

Algebraic inversion then gives values in $\{Z\}$ as

$$\{Z\} = [i\Omega[1] - [a]]^{-1}\{F_0\} \qquad (18\text{-}47)$$

A typical variable in the solution, y_k, is given in terms of the above by

$$y_k = \mathrm{Re}(Z_k e^{i\Omega t})$$
$$= Y_k \cos (\Omega t + \phi_k) \qquad (18\text{-}48)$$

in which

$$Y_k = |Z_k| \qquad (18\text{-}49)$$

$$\phi_k = \tan^{-1} \frac{\mathrm{Im}Z_k}{\mathrm{Re}Z_k} \qquad (18\text{-}50)$$

As an example, consider the response of the electromechanical shaker system to the sinusoidally varying emf $E(t) = E_0 \cos \Omega t$. With y measured from the static equilibrium position, the equations of motion (18-3) become

$$L\frac{di}{dt} + Ri + K\frac{dy}{dt} = E_0 \cos \Omega t$$

$$m\frac{d^2y}{dt^2} + ky - Ki = 0$$

We let

$$\left\{ \begin{array}{c} i(t) \\ y(t) \end{array} \right\} = \mathrm{Re}\left\{ \begin{array}{c} z_1(t) \\ z_2(t) \end{array} \right\}$$

and note that z_1 and z_2 must satisfy the equations

$$L\frac{dz_1}{dt} + Rz_1 + K\frac{dz_2}{dt} = E_0\, e^{i\Omega t}$$

$$m\frac{d^2z_2}{dt^2} + kz_2 - Kz_1 = 0$$

A solution to these will have the form

$$\left\{ \begin{array}{c} z_1 \\ z_2 \end{array} \right\} = \left\{ \begin{array}{c} Z_1 \\ Z_2 \end{array} \right\} e^{i\Omega t}$$

Substitution into the differential equations leads to

$$
\begin{bmatrix}
(i\Omega L + R) & i\Omega K \\
-K & (k - m\Omega^2)
\end{bmatrix}
\begin{Bmatrix}
Z_1 \\
Z_2
\end{Bmatrix}
=
\begin{Bmatrix}
E_0 \\
0
\end{Bmatrix}
$$

which may be inverted to yield the constants Z_1 and Z_2:

$$
Z_1 = \frac{E_0}{R + i\Omega \left(L + \dfrac{K^2}{k - m\Omega^2} \right)}
$$

$$
Z_2 = \frac{KE_0}{(k - m\Omega^2)R + i\Omega[(k - m\Omega^2)L + K^2]}
$$

A particular solution may now be written

$$
i(t) = I \cos (\Omega t - \phi_1)
$$
$$
y(t) = Y \cos (\Omega t - \phi_2)
$$

In which the amplitudes and phases are given by

$$
I = \frac{E_0}{\sqrt{R^2 + \Omega^2 \left(L + \dfrac{K^2}{k - m\Omega^2} \right)^2}}
$$

$$
\phi_1 = \tan^{-1} \left[\frac{\Omega}{R} \left(L + \frac{K^2}{k - m\Omega^2} \right) \right]
$$

$$
Y = \frac{KE_0}{\sqrt{(k - m\Omega^2)^2 R^2 + \Omega^2[(k - m\Omega^2)L + K^2]^2}}
$$

$$
\phi_2 = \phi_1
$$

Input and output.   Our mathematical models often contain more detail than is of practical interest; that is, many of the variables are of only secondary interest, have been introduced primarily to complete the mathematical model. For example, the current i in the electromechanical shaker system governed by Equation 18-3 would normally be of much less interest than the displacement y or the acceleration $\ddot{y}$. Or, the bending moment at the support section of the system of Figure 18-8 might be more interesting than the variables y_1 and y_2 of Equations 18-4. Thus, for a specific study we are led to define a quantity that we call the *output*, or *response*, $r(t)$. If the displacement of the shaker table is of interest, we define the output in terms of the system variables as

$$
r(t) = y = \begin{bmatrix} 0 & 1 \end{bmatrix} \begin{Bmatrix} i \\ y \end{Bmatrix}
$$

If the output is to be the bending moment at the support in Figure 18-8, then we similarly define

$$r(t) = f_1 a + f_2(2a)$$

$$= \left(\frac{96EI}{7a^3} y_1 - \frac{30EI}{7a^3} y_2\right)a + \left(-\frac{30EI}{7a^3} y_1 + \frac{12EI}{7a^3} y_2\right)2a$$

$$= \frac{6EI}{7a^2} [6 \quad -1] \left\{\begin{matrix} y_1 \\ y_2 \end{matrix}\right\}$$

In each case, the response is some linear combination of the variables introduced for modeling, so that we can write it as

$$r(t) = \{R\}^T\{y\} \tag{18-51}$$

This concept can be generalized to include a *set* of quantities as constituting the response, but here we will consider response to mean some single quantity, depending linearly on the system variables.

At the "other end" of the system are the sources of excitation which set the system in motion. In the electromechanical shaker, it is the emf applied to the driving coil that causes motion; thus, we call this the *input*, or *signal* $s(t) = E(t)$. Again, although the concept can be generalized to include a *set* of excitation sources, we restrict the definition here to a single quantity. In the mathematical model, the input s will appear in the forcing function, in the form

$$\{F(t)\} = \{F_0\} + \{S\}s(t) \tag{18-52}$$

For example, the equations for the shaker system can be written as

$$\begin{bmatrix} \left(L\dfrac{d}{dt} + R\right) & K\dfrac{d}{dt} \\ -K & \left(m\dfrac{d^2}{dt^2} + k\right) \end{bmatrix} \left\{\begin{matrix} i \\ y \end{matrix}\right\} = \left\{\begin{matrix} 0 \\ mg \end{matrix}\right\} + \left\{\begin{matrix} 1 \\ 0 \end{matrix}\right\}s(t)$$

The constant particular solution $\{y_0\}$, corresponding to the constant portion $\{F_0\}$ of the forcing function, will not be considered as part of the input-output relationship; these static values are most conveniently superimposed after the response to $s(t)$ is determined.

In the following sections we will see how to predict the responses to some particularly important signal forms, and how these may be used to predict responses to other forms.

Steady-state Frequency response. When the signal is sinusoidal in form,

$$s(t) = \cos \Omega t$$

the linear, constant coefficient system always possesses a solution in which every variable oscillates sinusoidally at the frequency of the signal. When so excited, systems in which all characteristic roots have negative real parts will ultimately reach this state of motion, even if initial conditions impart additional motions.

With the sinusoidal input of unit magnitude, the sinusoidal response,

$$r(t) = R \cos (\Omega t - \phi)$$

is called the *steady-state frequency response*. The quantity R, the ratio of output magnitude to input magnitude, is called the *gain*. The quantity ϕ is called the *phase lag*.

A procedure for finding the solution,

$$\{y(t)\} = \text{Re} \{Ze^{i\Omega t}\}$$

was outlined earlier. The response may be calculated from its definition in terms of $\{y\}$:

$$r(t) = \text{Re} (\{R\}^T \{Ze^{i\Omega t}\})$$

As an example, consider the system of Figure 18-3. Let the input be the vertical distance of the mass center of the rotor above the axis of rotation. Let the output be the force transmitted to the foundation (minus the gravitational force). The equation of motion is Equation 18-2:

$$m\ddot{y} + c\dot{y} + ky = -m_r\ddot{s}(t)$$

and the response is given in terms of the solution by

$$r = \left(k + c \frac{d}{dt} \right) y$$

With $s(t) = \text{Re } e^{i\Omega t}$ the steady-state solution can be determined as explained earlier, and is

$$y = \text{Re } z$$

$$z = \frac{m_r\Omega^2 \, e^{i\Omega t}}{(k - m\Omega^2) + i\Omega}$$

the response is then

$$r = \text{Re } (kz + c\dot{z})$$

$$= \text{Re } \frac{(k + ic\Omega)m_r\Omega^2 \, e^{i\Omega t}}{(k - m\Omega^2) + i(c\Omega)}$$

$$= \sqrt{\frac{1 + \dfrac{2\zeta\Omega}{\omega_0}}{\left(1 - \dfrac{\Omega^2}{\omega_0{}^2}\right)^2 + \left(\dfrac{2\zeta\Omega}{\omega_0}\right)^2}} \; \frac{\Omega^2}{\omega_0{}^2} \frac{km_r}{m} \cos(\Omega t - \phi)$$

where $\zeta = (c/2\sqrt{mk})$ and $\omega_0 = \sqrt{k/m}$ are the damping ratio and undamped natural frequency, respectively, and the phase lag is given by

$$\phi = \tan^{-1}\left(\frac{c\Omega}{k - m\Omega^2}\right) - \tan^{-1}\left(\frac{c\Omega}{k}\right)$$

$$= \tan^{-1}\left(\frac{2\zeta \dfrac{\Omega^3}{\omega_0{}^3}}{1 - (1 - 4\zeta^2)\dfrac{\Omega^2}{\omega_0{}^2}}\right)$$

Figure 18-15 shows the dependence of gain and phase lag on frequency and damping ratios. Perhaps the most important aspect of this is the *resonance phenomenon*. That is, as the driving frequency approaches the natural frequency of the system, very large gain will occur, unless the system is heavily damped.

Indicial response. The *unit step function* is defined as

$$u(t) = \begin{cases} 0 & t < 0 \\ 1 & t > 0 \end{cases} \tag{18-53}$$

When a linear system is initially at rest (i.e., all variables and their derivatives have the value zero when $t = 0$), and the input is the unit step function, the response is called the *indicial response*. We will denote this special response by $g(t)$. A surprising amount of this information about the system may be found from the indicial response; for example, we will show later how this response may be used to predict the response to an *arbitrary* input.

For illustration, consider the system of Figure 18-1, modeled by the first-order equation

$$\dot{y} + By = Cx \tag{18-1}$$

in which the control stick displacement $x(t)$ is the input and the actuator rod displacement $y(t)$ is the output. The indicial response then satisfies

$$\frac{dg}{dt} + Bg = Cu(t)$$

$$g(0) = 0$$

Figure 18-15

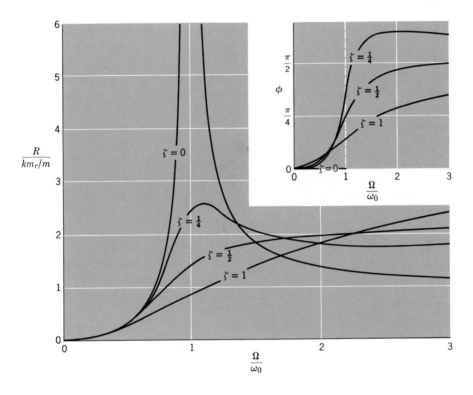

and may be determined readily by methods discussed earlier, to be

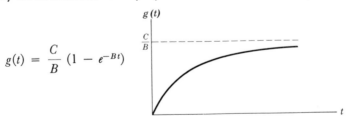

$$g(t) = \frac{C}{B} (1 - e^{-Bt})$$

This shows the manner in which the actuator displacement follows a sudden shift in control-stick displacement.

As a second example, consider the electromechankcal shaker system of Figure 18-5. Let the input be the impressed emf $E(t)$ and the output be the displacement measured from the equilibrium position, $y + (mg/k)$. The indicial response g then satisfies

$$L \frac{di}{dt} + Ri + K \frac{dg}{dt} = u(t)$$

$$m \frac{d^2g}{dt^2} + kg - Ki = 0$$

$$g(0) = \dot{g}(0) = i(0) = 0$$

The general solution to the differential equations is

$$\left\{ \begin{array}{c} \dfrac{Li}{K} \\ \\ g \end{array} \right\} = \left\{ \begin{array}{c} \dfrac{L}{KR} \\ \\ \dfrac{K}{kR} \end{array} \right\} + \sum_{n=1}^{3} C_n \left\{ \begin{array}{c} 1 \\ \\ \left(\dfrac{Y}{X}\right)_n \end{array} \right\} e^{\lambda_n t}$$

The three constants of integration may be determined from the initial conditions:

$$C_1 + C_2 + C_3 = - \frac{L}{KR}$$

$$\left(\frac{Y}{X}\right)_1 C_1 + \left(\frac{Y}{X}\right)_2 C_2 + \left(\frac{Y}{X}\right)_3 C_3 = - \frac{K}{kR}$$

$$\lambda_1 \left(\frac{Y}{X}\right)_1 C_1 + \lambda_2 \left(\frac{Y}{X}\right)_2 C_2 + \lambda_3 \left(\frac{Y}{X}\right)_3 C_3 = 0$$

For the example values of the parameters following Equation 18-35, these constants are

$$C_1 = -3.458 \ \mu\text{m/V}$$
$$C_{2,3} = (0.873 \ \mu\text{m/V}) \ e^{\mp 1.5704i}$$

The indicial response is then

$$g(t) = \Big[7.8242 - 7.8250 \ e^{-0.000 \ 442(Rt/L)}$$

$$+ \ 0.0017 \ e^{-0.4998(Rt/L)} \cos \left(1.979 \frac{Rt}{L} + 1.076 \right) \Big] \text{mm/V}$$

This is shown in Figure 18-16.

   Impulse response. The unit step function is a member of a family of *singularity functions*, which are characterized by their sharp fluctuations at the origin. A mathematically rigorous analysis of these presents a significant challenge*;

* A rigorous foundation for mathematical operations with the singularity functions was developed by L. Schwartz, *Théorie des Distributions*, Tome I et II, Actualitiés Scientifiques et Industrielles 1091 and 1122, Hermann & Cie, Paris, 1950, 1951.

Figure 18-16

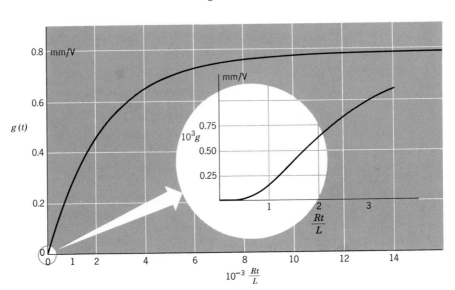

however, an intuitively plausible analysis that leads to correct results for our applications presents relatively little conceptual difficulty.

We consider the fluctuations in the functions to occur during a small, but finite, time increment. We can further prescribe some specific form for this fluctuation during the increment, although for our present purpose the important aspect is the total change that accrues rather than the form of the fluctuation. Figure 18-17 shows the unit step function as rising from 0 to 1 during the time increment 2ϵ, in a parabolic fashion. The derivative of the unit step function, called the *unit impulse* function and denoted by $\delta(t)$, then appears as triangular in form. The important properties of $\delta(t)$ are the relatively short duration in which it is nonzero, and the correspondingly high values it must reach such that the area under the curve in the $\delta - t$ plane has unit value.

The derivative of the unit impulse function is called the unit doublet function. Higher order singularity functions may be obtained by successive differentiation, producing functions with still more "wild" behavior near the origin.

Basic properties of the impulse function that are important for analysis of dynamic systems, are as follows:

1. If $f(t)$ is an "ordinary" function (i.e., its fluctuations are much slower than those of the singularity functions), the multiplication by the impulse function, followed by the integration, yields

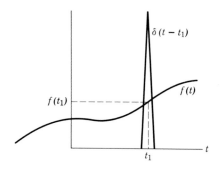

$$\int_{t_a}^{t_b} f(t)\delta(t - t_1)dt = f(t_1)u(t_1 - t_a)u(t_b - t_1)\qquad(18\text{-}54)$$

2. Consider a function $y^{(n)}(t)$, which contains an impulse at $t = t_1$ but it otherwise "ordinary," as depicted in Figure 18-18. The integral of this function, $y^{(n-1)}(t) = \int y^{(n)}(t)\,dt$, will have a discontinuity at $t = t_1$, and the integral of this function, $y^{(n-2)} = \int y^{(n-1)}\,dt$, as well as all successive integrals will be continuous. With this in mind, consider the differential equation

$$a_n\frac{d^n y}{dt^n} + a_{n-1}\frac{d^{n-1}y}{dt^{n-1}} + \cdots + a_0 y = \delta(t - t_1)\qquad(18\text{-}55a)$$

Since the general solution to the reduced equation and all its derivatives are "ordinary," the solution to this equation will exhibit singular behavior only near $t = t_1$. In view of the above, the lead term, $a_n(d^n y/dt^n)$, must contain an impulse singularity and the second term, $a_{n-1}(d^{n-1}y/dt^{n-1})$, a jump singularity at $t = t_1$. All lower derivatives will be continuous. Now let us integrate each term over the short interval $(t_1 - \epsilon, t_1 + \epsilon)$. With the abbreviation $t_1\pm$ for $t_1 \pm \epsilon$, the result may be written as

$$a_n\frac{d^{n-1}y}{dt^{n-1}}\bigg|_{t_1-}^{t_1+} + a_{n-1}\frac{d^{n-2}y}{dt^{n-2}}\bigg|_{t_1-}^{t_1+} + \cdots + a_0\int_{t_1-}^{t_1+} y\,dt = \int_{t_1-}^{t_1+}\delta(t - t_1)\,dt$$

Because all derivatives of y lower than the $n - 1$th are continuous, this reduces to

$$a_n[y^{(n-1)}(t_1+) - y^{(n-1)}(t_1-)] = 1\qquad(18\text{-}55b)$$

That is, the unit impulse produces a jump equal to $1/a_n$ in the $n - 1$th derivative of y.

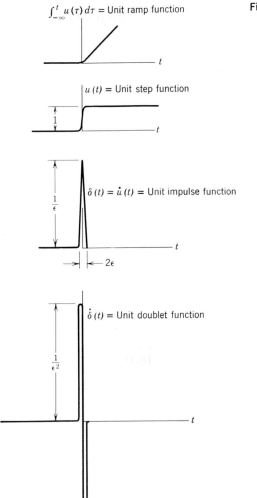

$\int_{-\infty}^{t} u(\tau)\, d\tau$ = Unit ramp function

Figure 18-17

$u(t)$ = Unit step function

$\delta(t) = \dot{u}(t)$ = Unit impulse function

$\dot{\delta}(t)$ = Unit doublet function

The last observation provides a means of constructing solutions to the above, impulse-forced system, using procedures already discussed. Suppose, for example, we wish to construct the solution to the equations

$$a_n \frac{d^n y}{dt^n} + a_{n-1} \frac{d^{n-1} y}{dt^{n-1}} + \cdots + a_0 y = \delta(t)$$

$$y^{(n-1)}(0-) = y^{(n-2)}(0-) = \cdots = y(0-) = 0$$

$y^{(n-2)} = \int y^{(n-1)}\, dt$ **Figure 18-18**

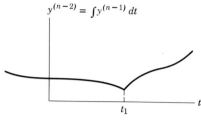

$y^{(n-1)} = \int y^{(n)}\, dt$

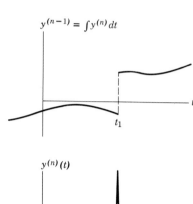

$y^{(n)}(t)$

which govern from time $t = 0-$ forward. With the knowledge of the effect of the impulse on y and its derivatives during the interval $(0-, 0+)$, we arrive at the following equations, which are equivalent from $t = 0+$ forward:

$$a_n \frac{d^n y}{dt^n} + a_{n-1} \frac{d^{n-1}y}{dt^{n-1}} + \cdots + a_0 y = 0$$

$$y^{(n-1)}(0+) = \frac{1}{a_n} \qquad y^{(n-2)}(0+) = \cdots = y(0+) = 0$$

Therefore, the solution will be of the form

$$y = C_1 e^{\lambda_1 t} + C_2 e^{\lambda_2 t} + \cdots + C_n e^{\lambda_n t}$$

with the constants of integration evaluated to satisfy the latter set of initial conditions.

The *impulse response* of a system is defined as the output resulting from a unit impulse input, with the system at rest initially. We shall denote this special response by $h(t)$. As an example, consider the spring-mass-dashpot system, governed by

$$m\ddot{y} + c\dot{y} + ky = F(t)$$

Consider as input the force $F(t)$ and as output the displacement of the mass. The impulse response then satisfies

$$m\ddot{h} + c\dot{h} + kh = \delta(t)$$
$$h(0-) = \dot{h}(0-) = 0$$

These equations characterize the effect on the suspended mass of a high-intensity, short-duration force. According to the above, this impulsive force will change the initial rest state of the system to the following:

$$h(0+) = 0$$
$$\dot{h}(0+) = \frac{1}{m}$$

That is, the impulsive force imparts a velocity to the mass, without altering its position during the short duration of the force, an effect that a sharp blow with a hammer might have.

A reference to Equations 18-29 and 18-31 gives the response after time $t - 0+$, as

$$h(t) = \frac{1}{m} e^{-ct/2m} \begin{cases} \dfrac{\sin \omega t}{\omega} & \dfrac{c}{2\sqrt{mk}} < 1 \\[2ex] \dfrac{\sinh \beta t}{\beta} & \dfrac{c}{2\sqrt{mk}} > 1 \end{cases}$$

Superposition integrals. An arbitrary forcing function or input signal may be constructed as a superposition of appropriately scaled and timed step functions, or appropriately scaled and timed impulse functions. Then, because of the linearity of the system, the response may be determined by means of superposition of indicial or impulse responses.

Figure 18-19

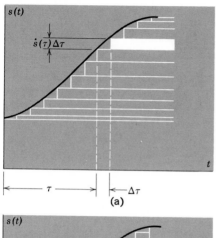

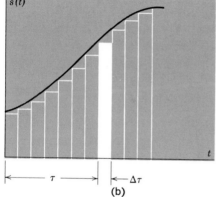

If the system has constant parameters in addition to being linear, the response following the arrival of the discontinuity in signal will not depend on the time of this arrival. That is, the responses to

$$s(t) = s_0 u(t - t_1)$$

and

$$s(t) = s_0 \delta(t - t_1)$$

will be

$$r(t) = s_0 g(t - t_1) \qquad t > t_1$$

and

$$r(t) = s_0 h(t - t_1) \qquad t > t_1$$

respectively.

In Figure 18-19a is shown an input $s(t)$, depicted as a sum of a number of step functions. The portion that is shaded has magnitude $\dot{s}(\tau)\Delta t$ and begins at time $t = \tau + \Delta\tau \simeq \tau$. Now the response of a linear, constant parameter system to the input

$$\dot{s}(\tau)\,\Delta\tau\,u(t - \tau)$$

would be

$$\Delta r = \dot{s}(\tau)\,\Delta\tau\,g(t - \tau) \qquad t > \tau$$

Superposing the responses to all such step inputs shown in the diagram, we have

$$\boxed{r(t) = s(0)g(t) + \int_0^t \dot{s}(\tau)g(t - \tau)\,d\tau} \qquad (18\text{-}56)$$

which may be used to obtain the response to an arbitrary signal.

Figure 18-19b depicts the input as made up of a series of impulse functions. The magnitude of the shaded impulse is $s(\tau)\Delta\tau$. The response at time t, induced by the impulse input,

$$s(\tau)\,\Delta\tau\,\delta(t - \tau)$$

would be

$$\Delta r = s(\tau)\,\Delta\tau\,h(t - \tau)$$

Superimposing the responses to all the impulse inputs shown in the diagram, we have

$$\boxed{r(t) = \int_0^t s(\tau)h(t - \tau)\,d\tau} \qquad (18\text{-}57)$$

which may also be used to obtain the response to an arbitrary signal.

Example

Determine the displacement y of an undamped spring-mass system governed by

$$m\ddot{y} + ky = F(t)$$

to the forcing function

$$F(t) = F_0\,e^{-bt}\,u(t)$$

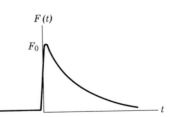

The indicial and impulse responses for this case are

$$g(t) = \frac{1}{k} (1 - \cos \omega_0 t)$$

and

$$h(t) = \frac{1}{m\omega_0} \sin \omega_0 t$$

respectively. Substituting into (18-56), we write

$$y(t) = \frac{F_0}{k} (1 - \cos \omega_0 t) + \int_0^t -F_0 b \; e^{-b\tau} \frac{1}{k} [1 - \cos \omega_0(t - \tau)] \, d\tau$$

As an alternative, we can substitute into (18-57):

$$y(t) = \int_0^t F_0 \, e^{-b\tau} \frac{1}{m\omega_0} \sin \omega_0(t - \tau) \, d\tau$$

Evaluation of either of these integrals leads to the response

$$y(t) = \frac{F_0}{k(1 + b^2/\omega_0^2)} \left(e^{-bt} - \cos \omega_0 t + \frac{b}{\omega_0} \sin \omega_0 t \right)$$

The above superposition integrals have the form

$$H(x) = \int_0^x F(\xi)G(x - \xi) \, d\xi \tag{18-58}$$

This function, often abbreviated as

$$H = F * G$$

is called the *convolution* of the functions $F(x)$ and $G(x)$. Evaluation of the convolution can sometimes be simplified by taking advantage of the symmetry of the operation,

$$F * G = \int_0^x F(\xi)G(x - \xi) \, d\xi = \int_0^x G(\xi)F(x - \xi) \, d\xi = G * F \tag{18-59}$$

Another relationship that is sometimes useful for handling these integrals is the formula

$$\frac{d}{dx} \int_{a(x)}^{b(x)} \phi(x,\xi) \, d\xi = \int_a^b \frac{\partial \phi}{\partial x} \, d\xi + \phi(x,b) \frac{db}{dx} - \phi(x,a) \frac{da}{dx} \tag{18-60}$$

Problems

18-68   In the system shown, let the given displacement a be sinusoidal: $a(t) = a_0 \cos \Omega t$. Determine the steady-state sinusoidal displacement of the suspended mass.

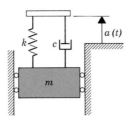

18-69   For the electromechanical shaker system, consider the emf applied to the coil as input and the table displacement as output. Plot a dimensionless measure of gain and the phase lag against a dimensionless frequency ratio, using values of the parameters given in the calculation of natural motions.

18-70   With $a(t)$ considered as input and the extension of k_0 considered as output, evaluate the steady-state frequency response of the system of Problem 18-4.

18-71   Suppose the structure of Figure 18-8 is excited as indicated. Determine the steady-state sinusoidally varying bending moment at the cantilever attachment.

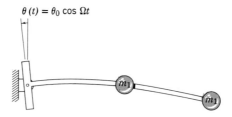

18-72   Determine the steady-state frequency response for the following systems:
(a) The accelerometer of Problem 18-5, with y as input and x as output.
(b) The accelerometer of Problem 18-5, with y as input and the force transmitted from m to the case as output.

18-73   Show, using sketches of curves of $f(t)$, $u(t - t_1)$, and $f(t)u(t - t_1)$, that

$$\frac{d}{dt} \left[f(t)u(t - t_1) \right] = f(t_1)\delta(t - t_1) + \dot{f}(t)u(t - t_1)$$

18-74   Determine the indicial response of the system of Problem 18-70, with the same definitions of input and output.

18-75   Determine the indicial response of the system of Problem 18-4, with $a(t)$ considered as input and the displacement of m considered as output.

18-76   With $f(t)$ considered as input and the force transmitted to the foundation as output, evaluate the steady-state frequency response, the impulse response, and the indicial response. Use the impulse response in (18-57) to obtain the indical response.

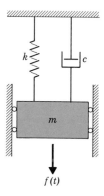

18-77   Integrate Equation 18-56 by parts and compare with Equation 18-57.

18-78   Using Equation 18-57, write an expression for the displacement of the block in the system of Problem 18-76 in terms of $f(t)$. By substitution into the differential equation of motion, verify this solution.

18-6

SOME PHENOMENA PECULIAR TO NONLINEAR AND PARA-METRICALLY EXCITED SYSTEMS.   Differential equations with time-varying parameters and/or nonlinearities cannot be integrated by the procedure described in the preceding section. Although the superposition theorems described in Section 18-4 apply to linear, time-varying systems, the construction of fundamental solutions to be used as bases for superposition is much more difficult than for constant-parameter systems. Considerable progress has been made in developing methods for constructing solutions; however, much remains to be done.

We will not describe the methods presently available for this task here, except to point out that such solutions are often approximations of one kind or another. Instead, we will describe a few of the more important phenomena that are peculiar to some of the dynamic systems that fall outside the linear, constant-parameter class.

Multiple Equilibrium Points.   A linear, autonomous system normally has just one set of values of the variables satisfying equilibrium. An exception occurs when the parameters are "critical" with respect to stability of the equilibrium, in which case the system can be in equilibrium throughout an entire subregion of the state space.

Unlike linear systems, nonlinear systems may contain many isolated equilibrium points—an example is the mechanism shown in Figure 18-12. This system can have two, three, or four distinct equilibrium points indicated by the roots of the nonlinear torque function $h(\phi)$ shown in Figure 18-13. Another example is the torque-free spinning of a rigid body, governed by the nonlinear Equations a, p. 177. These equilibrium states are represented by the points where the angular velocity resultant coincides with one of the three principal axes of inertia. The points, and their conditions of stability, are shown in Figure 16-11b.

Self-Excited Oscillations.   The *Van der Pol oscillator*, a system described by the differential equation

$$\frac{d^2x}{d\tau^2} - \alpha(1 - x^2)\frac{dx}{d\tau} + x = 0$$

is the most famous system exhibiting self-excited oscillation. Certain electronic circuits are modeled by this equation. Also, clutches and other mechanical devices that exhibit "chattering" are modeled by equations that, because of forces that depend nonlinearly on velocity, show the same features as Van der Pol's equation.

We can see from the equation that for x near the equilibrium point $x = 0$, the system will behave as a spring-mass combination with "negative damping", solutions to the linearized equations having the form

$$x = C\,e^{(\alpha/2)\tau}\cos(\omega\tau + \theta) \qquad \omega = \sqrt{1 - \frac{\alpha^2}{4}}$$

Therefore, any small disturbance will tend to diverge. However, after x increases, the nonlinear contribution becomes significant and tends to limit the amplitude of the oscillation. The oscillator soon reaches a *limit cycle*, in which the motion is periodic with an amplitude that depends on α, and not on the initial conditions. This behavior is illustrated in Figure 18-20.

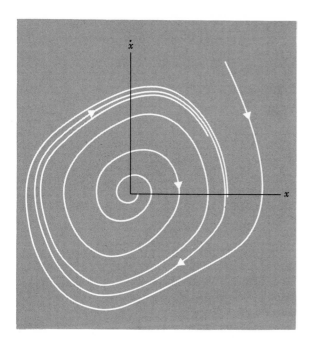

Figure 18-20

Jump Phenomenon.   Under the influence of a periodic excitation, a system with a nonlinear restoring force may be expected to exhibit a steady-state oscillatory response having the same frequency as that of the excitation, just as will a linear system. Furthermore, if the system is lightly damped, the amplitude of this response is observed to increase markedly when the frequency of the excitation approaches the natural frequency of the system—again much like the linear system.

However, the resonance diagram has a feature that is absent from the resonance diagram for the linear system. The "nose" is bent to one side, as indicated in Figure 18-21. With this feature goes the fact that there is a certain range of frequencies for which there exist three different steady-state periodic solutions, each with a different amplitude.

The periodic motion with the intermediate amplitude is unstable, and so is not actually observed in such systems. If the excitation frequency is slowly increased from below the resonance region, the amplitude of the response is observed to follow the upper curve through the region of multiple amplitudes. Any increase in frequency beyond the "nose" A of the curve is accompanied by a jump in the response, down to the lower branch. Similarly, if the frequency is slowly decreased from above, the response amplitude jumps to the upper

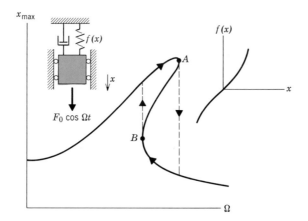

Figure 18-21

branch when the operation reaches point B on the diagram. This behavior is called the *jump phenomenon*.

Systems with Periodic Parametric Excitation.   *Mathieu's equation,*

$$\frac{d^2x}{d\tau^2} + (\alpha + 2\beta \cos 2\tau)x = 0$$

is the most famous model for periodic parametric excitation. For small angles, the pendulum with the vertically oscillating support, Problem 17-17, is described by this equation. Also, the small deviations from the steady-state response described in the preceding are well modeled by an equation of this type. Several other interesting examples are described in Den Hartog's *Mechanical Vibrations.* [*]

The equilibrium point $x = 0$ may be stable or unstable, depending on the values of the two parameters α and β. Moreover, these stabilities and instabilities defy intuition for the most part.

As can be determined from the chart in Figure 18-22, the equilibrium is unstable for small values of the excitation amplitude β, when the ratio of the excitation frequency to the natural frequency of the unexcited system is near one of the values 2/1, 2/2, 2/3, 2/4 . . . .

A more surprising result is the stable region to the left of the origin. In terms of the pendulum example cited, negative α corresponds to the inverted equilibrium position $\phi = \pi$, Figure 12-5. Thus there are certain combinations of frequency and amplitude that will *stabilize* this equilibrium position. Mechanical models of this have been built, and are truly amazing in this mode of operation.

---

[*] J. P. Den Hartog, *Mechanical Vibrations*, McGraw-Hill, 1956.

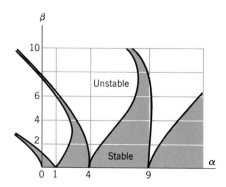

Figure 18-22

Problems

18-79   Identify the parameters of the system of Problem 17-7, for small oscillations, with the parameters α and β in Mathieu's equation. Design such a system that will exhibit the strange upside-down stability.

18-80   Determine values of the parameter mg/kl in the system of Figure 18-12, for which there are four, three, and two equilibrium points. Sketch the system configuration for each equilibrium. Plot curves of constant mechanical energy in the plane of ϕ versus $\dot{\phi}/\sqrt{k/m}$ for each of the three cases.

APPENDIX A
SOME USEFUL
NUMERICAL
VALUES

A-1

PHYSICAL CONSTANTS

Universal gravitational constant	6.67×10^{-11} N$\cdot$m^2/kg^2
Mass of the earth	5.983×10^{24} kg
Speed of light	0.2998 Gm/s
Standard atmospheric pressure	101.325 kPa
Density of air (0°C, 1 atm)	1.29 kg/m^3
Density of water (4°C, 1 atm)	1.0000 Mg/m^3
Density of concrete	2.4 Mg/m^3
Density of steel	7.7-7.9 Mg/m^3

A-2

PREFIXES FOR SI UNITS

Exponents in higher-order derived units apply to prefixes as well as units. For example, $km^2 = (10^3 m)^2 = 10^6 m^2$

Prefix	SI Symbol	Multiplication Factor
tera	T	10^{12}
giga	G	10^9
mega	M	10^6
kilo	k	10^3
hecto*	h*	10^2
deka*	da*	10^1
deci*	d*	10^{-1}
centi*	c*	10^{-2}
milli	m	10^{-3}
micro	μ	10^{-6}
nano	n	10^{-9}
pico	p	10^{-12}
femto	f	10^{-15}
atto	a	10^{-18}

A-3

UNITS OF MEASUREMENT

Quantity	SI Unit	Other Units
length	meter (m)	inch = 25.400 mm foot = 0.304 800 m statute mile = 1.609 344 km nautical mile = 1.852 000 km fathom = 1.828 800 m
mass	kilogram (kg)	pound-mass = 0.453 592 37 kg slug = 14.593 903 kg grain = 64.798 910 mg
time	second (s)	minute = 60 s hour = 3600 s
temperature	kelvin (K)	degrees Fahrenheit $t_K = (t_F + 459.67)/1.8$ degrees Celsius $t_K = t_C + 273.15$
plane angle	radian (rad)	degree = $(\pi/180)$ rad minute = $(1/60)$ deg second = $(1/60)$ min
angular velocity	rad/s	rpm = $(\pi/30)$ rad/s

* Use discouraged.

Quantity	SI Unit	Other Units
area	m²	acre = 4046.856 m² hectare = 10⁴ m²
energy	joule (J) J = N·m = kg·m²/s²	foot-pound-force = 1.355 818 J erg = 10⁻⁷ J calorie (mean) = 4.190 02 J British thermal unit = 1.055 kJ
force	newton (N) N = kg·m/s²	pound-force = 4.448 222 N kip = 4.448 222 kN poundal = 0.138 255 N dyne = 10⁻⁵ N kilogram-force = 9.806 650 N
power	watt (W) W = J/s = kg·m²/s³	horsepower (550 ft-lbf/s) = 745.7 W British thermal unit per hour = 0.293 071 W refrigeration ton = 3.517 kW
pressure	pascal (Pa) Pa = N/m² = kg/m·s²	bar = 100 kPa pound per square inch = 6.894 757 kPa centimeter of mercury (0°C) = 1.333 22 kPa centimeter of water (4°C) = 98.0638 Pa
velocity	m/s	kilometer per hour = 0.277 778 m/s mile per hour = 0.447 040 m/s knot = 0.514 444 m/s
volume	m³	liter = dm³ = 10⁻³ m³ fluid ounce = 2.957 353 × 10⁻⁵ m³ U. S. liquid gallon = 3.785 412 × 10⁻³ m³ barrel = 0.158 987 m³
frequency	hertz (Hz)	cycle per second = 1 Hz
electric current	ampere (A)	
electric charge	coulomb (C) C = A·s	
electromotive force	volt (V) V = W/A	
inductance	henry (H) H = V·s/A	
electric resistance	ohm (Ω) Ω = V/A	
electric capacitance	farad (F) F = A·s/V	

APPENDIX B
SOME FORMULAS
OF VECTOR
ANALYSIS

B-1

DEFINITIONS

A vector **A** specifies magnitude and direction, but no more. In order to specify "line of action" or "point of application" of a vector (such as for a force vector), an additional vector—a position vector—is used.

Magnitude: $A = |\mathbf{A}|$

$\mathbf{A}_x$ = projection of **A** onto x

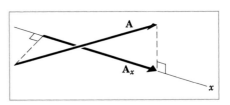

Addition: $\mathbf{A} + \mathbf{B}$

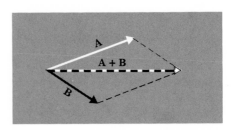

224

Negative: $-\mathbf{A}$

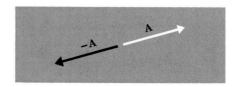

Subtraction: $\mathbf{A} - \mathbf{B} = \mathbf{A} + (-\mathbf{B})$

Multiplication by scalar:
$$|p\mathbf{A}| = |p||\mathbf{A}|$$

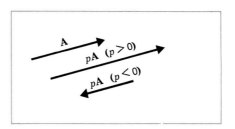

Dot product: $\mathbf{A} \cdot \mathbf{B} = AB \cos \sphericalangle\,{}^{\mathbf{B}}_{\mathbf{A}}$

Cross product:

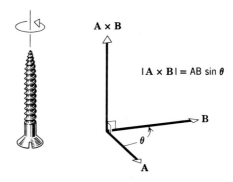

$|\mathbf{A} \times \mathbf{B}| = AB \sin \theta$

Derivative: $\overset{\alpha}{\mathbf{A}} = \underset{\Delta t \to 0}{\text{Lim}}\ \Delta_\alpha \mathbf{A}/\Delta t$
where $\Delta_\alpha \mathbf{A}$ is the change, observed in
the α reference frame, of the vector $\mathbf{A}$,
corresponding to the change in t, Δt.

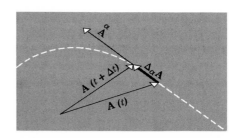

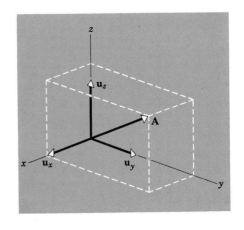

Rectangular Cartesian components:

$$\mathbf{A} = A_x\mathbf{u}_x + A_y\mathbf{u}_y + A_z\mathbf{u}_z$$

$$\mathbf{u}_i = \frac{\mathbf{A}_i}{A_i}$$

B-2

FORMULAS

$$\mathbf{A}\cdot\mathbf{B} = \mathbf{B}\cdot\mathbf{A} \qquad [3\text{-}7]$$

$$(\mathbf{A} + \mathbf{B})\cdot\mathbf{C} = \mathbf{A}\cdot\mathbf{C} + \mathbf{B}\cdot\mathbf{C} \qquad [3\text{-}8]$$

$$\mathbf{A} \times \mathbf{B} = -\mathbf{B} \times \mathbf{A} \qquad [3\text{-}10]$$

$$\mathbf{A} \times (\mathbf{B} + \mathbf{C}) = \mathbf{A} \times \mathbf{B} + \mathbf{A} \times \mathbf{C} \qquad [3\text{-}11]$$

$$\mathbf{A} \times (\mathbf{B} \times \mathbf{C}) = (\mathbf{C}\cdot\mathbf{A})\mathbf{B} - (\mathbf{A}\cdot\mathbf{B})\mathbf{C} \qquad [3\text{-}13]$$

$$\mathbf{A}\cdot(\mathbf{B} \times \mathbf{C}) = \mathbf{B}\cdot(\mathbf{C} \times \mathbf{A}) = \mathbf{C}\cdot(\mathbf{A} \times \mathbf{B}) \qquad [3\text{-}14]$$

$$\mathbf{A} = \frac{(\mathbf{B}\cdot\mathbf{A})\mathbf{B}}{\mathbf{B}\cdot\mathbf{B}} + \frac{(\mathbf{B} \times \mathbf{A}) \times \mathbf{B}}{\mathbf{B}\cdot\mathbf{B}} \qquad [3\text{-}15]$$

$$\mathbf{A}\cdot\mathbf{B} = A_xB_x + A_yB_y + A_zB_z \qquad [3\text{-}20]$$

$$\mathbf{A} \times \mathbf{B} = (A_yB_z - A_zB_y)\mathbf{u}_x + (A_zB_x - A_xB_z)\mathbf{u}_y + (A_xB_y - A_yB_x)\mathbf{u}_z \qquad [3\text{-}22]$$

$$\mathbf{A}\cdot(\mathbf{B} \times \mathbf{C}) = \begin{vmatrix} A_x & A_y & A_z \\ B_x & B_y & B_z \\ C_x & C_y & C_z \end{vmatrix} \qquad [3\text{-}23]$$

$$\overset{\alpha}{\mathbf{A}} = \dot{A}_x\mathbf{u}_x + \dot{A}_y\mathbf{u}_y + \dot{A}_z\mathbf{u}_z$$

provided the $\mathbf{u}_i$ are fixed in the reference frame α.

$$\overset{\alpha}{\widetilde{(p\mathbf{A})}} = \dot{p}\mathbf{A} + p\overset{\alpha}{\mathbf{A}} \qquad [9\text{-}6]$$

$$\overset{\alpha}{\widetilde{(\mathbf{A}\cdot\mathbf{B})}} = \overset{\alpha}{\mathbf{A}}\cdot\mathbf{B} + \mathbf{A}\cdot\overset{\alpha}{\mathbf{B}} \qquad [9\text{-}7]$$

$$\overset{\alpha}{\widetilde{(\mathbf{A} \times \mathbf{B})}} = \overset{\alpha}{\mathbf{A}} \times \mathbf{B} + \mathbf{A} \times \overset{\alpha}{\mathbf{B}} \qquad [9\text{-}8]$$

APPENDIX C
SOME FORMULAS
OF MATRIX ALGEBRA

C-1

DEFINITIONS

A matrix is an ordered, two-dimensional array of quantities obeying rules of algebra given below. A typical element in the matrix $[A]$ is written as A_{ij}, in which the indexes indicate that the element belongs in the ith row and the jth column. Subscripts i and j take on the integer values $i = 1, 2, 3, \cdots m; j = 1, 2, 3, \cdots n$.

Equality: $[A] = [B]$ implies that $A_{ij} = B_{ij}$ for all i and j.

Addition and Subtraction: $[C] = [A] \pm [B]$ implies that $C_{ij} = A_{ij} \pm B_{ij}$ for all i and j.

Multiplication: $[B] = \alpha[A]$ implies that $B_{ij} = \alpha A_{ij}$ for all i and j. $[C] = [A][B]$ implies that $C_{ij} = \sum_k A_{ik} B_{kj}$.

Transpose: $[B] = [A]^T$ implies that $B_{ij} = A_{ji}$.

Inverse: $[B] = [A]^{-1}$ implies that $[B][A] = [A][B] = [1]$

where

$$[1] = \begin{bmatrix} 1 & 0 & 0 & \cdots & 0 \\ 0 & 1 & 0 & \cdots & 0 \\ 0 & 0 & 1 & \cdots & 0 \\ \cdot & & & & \\ \cdot & & & & \\ \cdot & & & & \\ 0 & \cdots & 0 & 0 & 1 \end{bmatrix}$$

The inverse exists iff $\det[A] \neq 0$

The determinant of a square matrix, abbreviated as $\det[A]$, is the determinant obtained by considering the elements in the array as a determinant.

227

The adjoint of a square matrix, abbreviated as adj[A], is the transpose of the matrix obtained by replacing each element by its cofactor. The *cofactor* of the element A_{ij} is $(-1)^{i+j}$ times the determinant obtained by striking out the ith row and jth column.

C-2

FORMULAS

1.
$$[A] \pm [B] = [B] \pm [A]$$

2.
$$[A][[B] \pm [C]] = [A][B] \pm [A][C]$$
$$[[A] \pm [B]] [C] = [A][C] \pm [B][C]$$

3.
$$[A] [[B] [C]] = [[A] [B]] [C]$$

4.
$$[1] [A] = [A] [1] = [A]$$

5.
$$[A]^{-1} = \frac{\text{adj } [A]}{\det [A]}$$

6.
$$[[A] [B]]^{T} = [B]^{T}[A]^{T}$$

7.
$$[[A] [B]]^{-1} = [B]^{-1}[A]^{-1}$$

APPENDIX D
PROPERTIES
OF LINES, AREAS,
VOLUMES, AND
SOLIDS

D-1

LINES

$$L = \text{length} \qquad c_i = \frac{1}{L}\int i\, dL \qquad (i = x,y,z)$$

$$(c_x, c_y, c_z) = \text{coordinates of centroid}$$

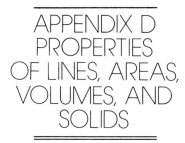

	Segment of circle	Parabola
$L =$	$2a\alpha$	$-\frac{a}{2}[u + \frac{a}{2b}\log(u + 2b/a)]$ $u = \sqrt{1 + (2b/a)^2}$
$c_x =$	$\dfrac{a \sin \alpha}{\alpha}$	$\dfrac{a^4}{12b^2L}(u^3 - 1)$ $u = \sqrt{1 + (2b/a)^2}$
$c_y =$	0	$\dfrac{a^3}{32\, b\, L}\{[1 + 2(2b/a)^2]u - (a/2b)\log(u + 2b/a)\}$ $u = \sqrt{1 + (2b/a)^2}$

229

D-2

PLANE AREAS

$$A = \text{area} \qquad c_i = \frac{1}{A} \int i \, dA \qquad (i = x,y)$$

$$A_{xx} = \int y^2 \, dA \qquad A_{yy} = \int x^2 \, dA \qquad A_{xy} = -\int xy \, dA$$

The second moments of area, denoted here as A_{ij}, were denoted as I_{ij} in *Statics*. In this volume, I_{ij} denotes second moment of mass.

$A =$	ab	$\dfrac{ab}{2}$	$\dfrac{ab}{n+1}$
$c_x =$	0	$\dfrac{a+c}{3}$	$\dfrac{n+1}{n+2}a$
$c_y =$	$\dfrac{b}{2}$	$\dfrac{b}{3}$	$\dfrac{(n+1)b}{2(2n+1)}$
$A_{xx} =$	$\dfrac{ab^3}{3}$	$\dfrac{ab^3}{12}$	$\dfrac{ab^3}{3(3n+1)}$
$A_{yy} =$	$\dfrac{a^3 b}{12}$	$\dfrac{ab}{12}(a^2 + ac + c^2)$	$\dfrac{a^3 b}{n+3}$
$A_{xy} =$	0	$-\dfrac{b^2 a}{24}(2c + a)$	$-\dfrac{a^2 b^2}{4(n+1)}$

	Quarter ellipse	Sector of circle	Segment of circle
$A =$	$\dfrac{\pi\,a\,b}{4}$	$a^2\,\alpha$	$a^2\left(\alpha - \dfrac{1}{2}\sin 2\alpha\right)$
c_x	$\dfrac{4\,a}{3\,\pi}$	$\dfrac{2a\sin\alpha}{3\,\alpha}$	$\dfrac{4a}{3}\dfrac{\sin^3\alpha}{2\alpha - \sin 2\alpha}$
$c_y =$	$\dfrac{4\,b}{3\,\pi}$	0	0
$A_{xx} =$	$\dfrac{\pi\,a\,b^3}{16}$	$\dfrac{a^4}{8}(2\alpha - \sin 2\alpha)$	$\dfrac{a^4}{8}\left(2\alpha - \dfrac{4}{3}\sin 2\alpha + \dfrac{1}{6}\sin 4\alpha\right)$
$A_{yy} =$	$\dfrac{\pi\,a^3\,b}{16}$	$\dfrac{a^4}{8}(2\alpha + \sin 2\alpha)$	$\dfrac{a^4}{8}\left(2\alpha - \dfrac{1}{2}\sin 4\alpha\right)$
$A_{xy} =$	$-\dfrac{b^2 a^2}{8}$	0	0

D-3

VOLUMES

$$V = \text{volume} \qquad c_i = \frac{1}{V} \int i \, dV \qquad (i = x,y,z)$$

$$(c_x, c_y, c_z) = \text{coordinates of centroid}$$

	Wedge	Segment of sphere
		$h = a\,(1 - \cos \alpha)$
$V =$	$\dfrac{a\,b\,c}{2}$	$\pi\,h^2\!\left(a - \dfrac{h}{3}\right)$
$c_x =$	$\dfrac{2\,a}{3}$	$a\,\dfrac{(1 - h/2a)^2}{(1 - h/3a)}$
$c_y =$	$\dfrac{b}{3}$	0
$c_z =$	$\dfrac{c}{2}$	0

Cone

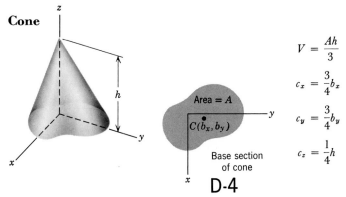

$$V = \frac{Ah}{3}$$

$$c_x = \frac{3}{4}b_x$$

$$c_y = \frac{3}{4}b_y$$

Area = A

$C(b_x, b_y)$

Base section
of cone

$$c_z = \frac{1}{4}h$$

D-4
SECOND MOMENTS OF MASS OF SOME HOMOGENEOUS BODIES

$$I_{xx} = \int_m (y^2 + z^2)dm \qquad I_{yy} = \int_m (z^2 + x^2)dm \qquad I_{zz} = \int_m (x^2 + y^2)dm$$

$$I_{xy} = -\int_m xy\, dm \qquad I_{yz} = -\int_m yz\, dm \qquad I_{zx} = -\int_m zx\, dm$$

Circular Segment of Slender Rod

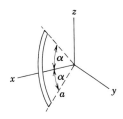

$$I_{xx} = \frac{ma^2}{2}\left(1 - \frac{\sin 2\alpha}{2\alpha}\right)$$

$$I_{yy} = ma^2$$

$$I_{zz} = \frac{ma^2}{2}\left(1 + \frac{\cos 2\alpha}{2\alpha}\right)$$

Straight Slender Rod

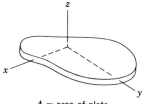

$$I_{XX} = I_{YY} = \frac{mL^2}{12}$$

Thin Flat Plate

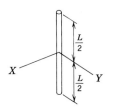

A = area of plate
A_{ij} = second moments
 of area

$$I_{ij} = \frac{mA_{ij}}{A} \qquad (i,j = x,y)$$

$$I_{zz} = I_{xx} + I_{yy}$$

Cylinder

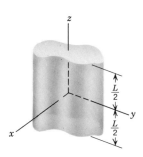

$$I_{xx} = m\left(\frac{A_{xx}}{A} + \frac{L^2}{12}\right)$$

$$I_{xy} = m\frac{A_{xy}}{A}$$

$$I_{yy} = m\left(\frac{A_{yy}}{A} + \frac{L^2}{12}\right)$$

$$I_{zz} = m\frac{A_{xx} + A_{yy}}{A}$$

A = area of cross section
A_{ij} = second moments of area of cross section

Cone

$$I_{xx} = m\left(\frac{3A_{xx}}{5A} + \frac{h^2}{10}\right)$$

$$I_{xy} = m\frac{3A_{xy}}{5A}$$

$$I_{xz} = -\frac{3mb_x h}{20}$$

$$I_{yy} = m\left(\frac{3A_{yy}}{5A} + \frac{h^2}{10}\right)$$

$$I_{yz} = -\frac{3mb_y h}{20}$$

$$I_{zz} = \frac{3m}{5}\left(\frac{A_{xx} + A_{yy}}{A}\right)$$

A = area of base section
(b_x, b_y) = coordinates of centroid of base section
A_{ij} = second moments of area of base section

Ellipsoid

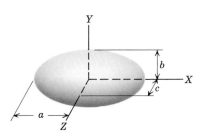

$$I_{XX} = \frac{1}{5}m(b^2 + c^2)$$

$$I_{YY} = \frac{1}{5}m(c^2 + a^2)$$

$$I_{ZZ} = \frac{1}{5}m(a^2 + b^2)$$

Segment of Sphere

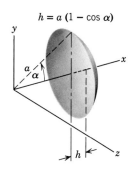

$$h = a\,(1 - \cos\alpha)$$

$$I_{xx} = \frac{2mah}{3}\left(\frac{1 - \dfrac{3h}{4a} + \dfrac{3h^2}{20a^2}}{1 - \dfrac{h}{3a}}\right)$$

$$I_{yy} = ma^2\left(\frac{1 - \dfrac{4h}{3a} + \dfrac{3h^2}{4a^2} - \dfrac{3h^3}{20a^3}}{1 - \dfrac{h}{3a}}\right)$$

APPENDIX E
SOME RELATIONSHIPS
AMONG COMPLEX
NUMBERS

DEFINITIONS

1. A complex number z is an ordered pair of real numbers (x,y), usually written as

$$z = x + iy$$

and satisfying the following algebraic rules:

$$az = ax + i\,ay \quad (a \text{ real})$$
$$z_1 \pm z_2 = (x_1 \pm x_2) + i(y_1 \pm y_2)$$
$$z_1 z_2 = (x_1 x_2 - y_1 y_2) + i(x_1 y_2 + x_1 y_2)$$
$$\frac{z_1}{z_2} = \frac{x_1 x_2 + y_1 y_2}{x_1^2 + y_2^2} + i\frac{x_2 y_1 - x_1 y_2}{x_2^2 + y_2^2}$$

These rules may be obtained formally by treating the quantities $(x_1 + iy_1)$ and $(x_2 + iy_2)$ in the same way as real binominals and setting $i^2 = -1$.
The real numbers x and y are called the real and imaginary parts, respectively, of $z = x + iy$. These are often written as

$$x = \text{Re } z \qquad y = \text{Im } z$$

2. The complex conjugate of $z = x + iy$ is defined as $x - iy$, written as

$$z^* = x - iy$$

Another common notation for z^* is $\bar{z}$.

236

3.   The *exponential* function is defined as

$$e^z = e^x(\cos y + i \sin y)$$

4. $$\frac{dz}{dt} = \frac{dx}{dt} + i\frac{dy}{dt} \qquad (t \text{ real})$$

USEFUL RELATIONSHIPS

The following relationships may be derived from the above definitions:

1. $x = \frac{1}{2}(z + z^*) \qquad y = \frac{1}{2i}(z - z^*)$

2. $(z^*)^* = z$

3. $(z_1 \pm z_2)^* = z_1{}^* \pm z_2{}^*$

4. $\left(\dfrac{z_1 z_2}{z_3}\right)^* = \dfrac{z_1{}^* z_2{}^*}{z_3{}^*}$

5. $(e^z)^* = e^{z^*}$

6. $e^{z_1}e^{z_2} = e^{(z_1 + z_2)}$

7. $z = re^{i\theta}$

  $z^* = re^{-i\theta}$

  where

  $r = \sqrt{x^2 + y^2}$

  $\theta = \tan^{-1}(y/x)$

8. $\dfrac{z_1 z_2}{z_3} = \dfrac{r_1 r_2}{r_3}e^{i(\theta_1 + \theta_2 - \theta_3)}$

9. $\cos\theta = \dfrac{e^{i\theta} + e^{-i\theta}}{2} \quad \sin\theta = \dfrac{e^{i\theta} + e^{-i\theta}}{2i}$

10. $\dfrac{d}{dt}(e^z) = e^z\dfrac{dz}{dt}$

11. If the combination $u = C_1 e^{\lambda t} + C_2 e^{\lambda^* t}$ is real for all t, then $C_2 = C_1{}^*$. Proof:

$$\text{Im}(C_1 e^{\lambda t} + C_2 e^{\lambda^* t}) = 0$$
$$(C_1 e^{\lambda t} + C_2 e^{\lambda^* t}) - (C_1 e^{\lambda t} + C_2 e^{\lambda^* t})^* = 0$$
$$(C_1 - C_2{}^*)e^{\lambda t} - (C_1 - C_2{}^*)^* e^{\lambda^* t} = 0$$
$$\text{Im}(C_1 - C_2{}^*)e^{\lambda t} = 0$$
$$\text{Call } C_1 - C_2{}^* = A + iB$$
$$\lambda = \alpha + i\omega$$
$$\text{Im}(C_1 - C_2{}^*)e^{\lambda t} = e^{\alpha t}(A \sin\omega t + B \cos\omega t)$$
$$\text{Let } t = 0: B = 0$$
$$\text{Let } t = \frac{\pi}{2\omega} : A = 0$$
$$C_1 - C_2{}^* = 0$$

REFERENCES

Isaac Asimov, *Asimov's Biographical Encyclopedia of Science and Technology* (Doubleday and Company, 1972).

Richard Bellman, *Introduction to Matrix Analysis*, Second Edition, (McGraw-Hill Book Company, 1970).

Stephen H. Crandall, D. C. Karnopp, E. F. Kurtz, and D. C. Pridmore-Brown, *Dynamics of Mechanical and Electromechanical Systems* (McGraw-Hill Book Company, 1968).

Herbert Goldstein, *Classical Mechanics* (Addison-Wesley Publishing Company, 1959).

Robert L. Halfman, *Dynamics*, Vols. I and II (Addison-Wesley Publishing Company, 1962).

Leonard Meirovitch, *Methods of Analytical Dynamics* (McGraw-Hill Book Company, 1970).

Edward J. Routh, *Dynamics of a System of Rigid Bodies, Part I*, Seventh Edition (Macmillan and Company, 1905).

Edward J. Routh, *Dynamics of a System of Rigid Bodies, Part II*, Sixth Edition (Macmillan and Company, 1905).

Murray R. Spiegel, *Vector Analysis* (Schaum's Outline Series, Schaum Publishing Company, 1959).

E. T. Whittaker, *A Treatise on the Analytical Dynamics of Particles and Rigid Bodies*, Fourth Edition (Cambridge University Press, 1961).

INDEX

Acceleration, 7
  differences, 16
  relationship for moving reference frames,
    9-10
Aerodynamic forces, 59
Amplitude ratio, 193
Angular acceleration, 38
Angular momentum, 68, 98, 108, 110
Angular velocity, 29, 45, 93
  composition of, 35
  relationship with linear velocities, 30
  in terms of linear velocities, 31-32
Antenna, 116
Autonomous system, 172
Auxiliary equations, 182
Axis
  coordinate, 57, 64-66
  instantaneous, 37, 40
  of inertial symmetry, 110, 129
  principal, 71, 103, 108
  screw, 40, 80
  of single equivalent rotation, 79, 83-86

Bevel gear train, 38
Body cone, 130

Cauchy's inertia ellipsoid, 107, 128
Centripetal component of acceleration, 11
Centroids, 229-232

Characteristic equation, 186, 187, 191,
    194
Chasle's theorem, 80
Complementary function, 183
Complex numbers, 236-237
Complex roots, 188, 194
Components of a vector, 1
Conformable matrices, 51
Conservative force, 144
Constant coefficient system, 172
Constants of integration, 187, 188, 194,
    195
Constraints, 135
Convolution, 213
Coordinate axis
  rotation, 57, 69-70
  translation, 102
Coriolis component of acceleration, 12,
    27
Coriolis, G. de, 13
Critical damping, 189

D'Alembert's principle, 149
Damping ratio, 187
Dashpot, 162
Degrees of freedom, 136
Dependent variables, 170-171
Derivatives of a tensor, 76
Derivatives of a vector, 1

in matrix notation, 63
Direct precession, 130
Direction cosines, 57
Distributed parameters, 171
Drag force, 59

Eigenvalues, 72, 103, 191
Eigenvectors, 72, 103, 197
Electromagnetic coupling, 163-164, 168
Electromechanical shaker, 163, 190, 199, 204
Ellipsoid of inertia, 107, 128
Elliptic integral, 156
Energy integral, 154
Equilibrium
  motion, 132, 179
  point, 176, 178
Euler's angles, 91, 94, 98, 130
Euler's equations of rigid body motion, 119
Euler's theorem on rigid rotations, 79
Exponential
  decay, 193, 195
  growth, 195

Flywheel, 36
Forcing functions, 172, 197, 199
Frequency of oscillation, 188, 193, 195
Frequency response, 198, 201
Friction forces, 146

Gain, 202
General solution, 184
  for first order l.c.c. system, 186
  for nth order l.c.c. system, 194-195
  for second order l.c.c. system, 187, 188, 189
Generalized coordinates, 136
Generalized force, 142
Generalized momentum, 154
Gimbal frame, 36
Gradient, 109
Gyrocompass, 122
Gyroscopic moment, 116

Hamiltonian function, 153
Helicopter rotor, 41

Holonomic systems, 136
Humpage's reduction gear train, 40
Hydraulic servo-actuator, 160

Identity matrix, 53
Ignorable coordinates, 154
Imaginary part of a complex number, 231
Impulse function, 206
Impulse response, 210
Independent variables, 171
Indicial response, 203
Inertia tensor for a rigid body, 71, 101
Infinite-degree-of-freedom structure, 166
Initial conditions, 181
Input, 201
Invariable plane, 127
Inverse of a matrix, 54, 227, 228
Inverse rotation of coordinate axes, 62
Instantaneous axis of rotation, 37, 126
Intrinsic coordinates, 16, 17, 55

Jump phenomenon, 217

Kane, T. R., 136
Kinetic energy, 150
  of a rigid body, 106, 124

Lagrange, J. L., 135
Lagrange's equations, 149-150
Lagrangian function, 152
l.c.c. system, 172
Lift force, 59
Limit cycle, 216
Linear independence, 183
Linear systems, 171, 175
Linearization, 175
Load-deflection relationships, 165
Lumped parameters, 171

Mathieu's equation, 218
Matrix, 49
  addition of, 50
  algebra, 49, 227-228
  associative law of multiplication, 61
  commutative law of multiplication, 52
  direction cosine, 58
  distributive law, 52, 55, 56

equality of, 50
inverse of, 54, 227, 228
multiplication of, 50
symmetric, 72
transpose of, 51
Moment-angular momentum relationship, 114
Moments of inertia, 100, 102, 109, 233-235
parallel axis formulas, 102
Momentum
angular, 68, 98, 101, 110
generalized, 154
linear, 101
Multiple equilibrium points, 216

Natural motions, 185
Newton's second law and moving reference frames, 24
Nonautonomous system, 172
Nonlinear systems, 171, 175, 215
Nutation, 132

Orthogonal transformations, 81
Orthonormal matrix, 62, 81
Oscillator, 162, 216
Oscillatory motion, 188, 195
Overdamped systems, 189

Parallel axis formulas, 102
Parameters, 170
distributed, 171
lumped, 171
Parametric excitation, 218
Parametrically excited system, 172, 218
Particular solution, 181, 182, 197
Pendulum, 24, 218
Phase lag, 202
Plane of symmetry, 100
Poinsot, L., 127
Poinsot mechanism, 127
Polar form of a complex number, 237
(item 7)
Polhode, 128
Position vector, 5
Postfactor, 52
Potential function, 145

Precession
direct, 130
retrograde, 130
secondary, 132
Prefactor, 52
Principal axes of inertia, 71, 103, 108
Principal directions, 71
Products of inertia, 100, 233-235
parallel axis formulas, 102

Real part of a complex number, 236
Reduced equations, 182
Reference frame, 1
effect on derivative of a vector, 4
Relative acceleration, 10
Relative amplitude, 193
Relative velocity, 10
Resonance, 203
Response, 185, 200
frequency, 201
impulse, 210
indicial, 203
Retrograde precession, 130
Rigid body
angular momentum, 68, 98, 108, 110
Chasle's theorem, 80
displacement kinematics, 77
Euler's angles, 91, 94, 98
Euler's theorem on rigid rotations, 79
inertially symmetric, 110
inertially spherical, 110
kinetic energy of, 106
moment of inertia of, 100, 102, 109, 233-235
product of inertia of, 100, 102, 233-235
screw displacement, 80
small rotations, 88
work done on, 124
Rigidity constraint, 30
Rotation of coordinate axes, 57, 69-70
Rotational displacement
Euler's theorem, 79
Infinitesimal, 88
Single, fixed axis, 79, 83
Routh, E. J., 155

Satellite, 7

Screw axis, 40
Screw displacement, 80
Screw motion, 40
Self-excited oscillations, 216
Shaker, electromechanical, 163, 190, 199,
    204
Signal, 201
Singularity functions, 205
Small rotations of a rigid body, 88
Solutions of differential equations, 181
    particular, 181
    general, 184
Space cone, 130
Spherical coordinates, 16
Spherical, inertially, 110
Stable motion, 129
State variable form, 172
Step function, 203
Stress at a point, 76
Structure
    an infinite-degree-of-freedom, 166
    a two-degree-of-freedom, 165
Superposition, 180
Superposition integrals, 210-212
Symmetric matrix, 72
Symmetry, plane of, 100
Systems, 159
    constant-coefficient, 159
    linear, 159

Taylor's series, 176, 179

Tensor, 70
    derivatives of, 116
Third order l.c.c. system, 177, 190
Top, spinning, 130, 155
Torque-free motion of a rigid body, 127
Transpose of a matrix, 52

Undamped systems, 189
Underdamped systems, 189
Unit matrix, 53
Universal joint, 32, 42

Van der Pol oscillator, 216
Variables, 167
    dependent, 170-171
    independent, 171
Vector
    addition, 224
    components of, 1
    cross product, 225, 226
    derivatives of, 1, 63, 225, 226
    dot product, 225, 226
    rectangular Cartesian components, 226
    subtraction, 225
    transformations, 67
Velocity, 5
    differences, 16
    relationship for moving reference frame
        9-10

Work-kinetic energy integral, 125

$$\overset{\alpha}{\mathbf{A}} = \overset{\beta}{\mathbf{A}} + {}_{\alpha}\Omega_{\beta} \times \mathbf{A}$$